INVERSE AND ILL-POSED PROBLEMS SERIES

Inverse Problems of Electromagnetic Geophysical Fields

Also available in the Inverse and Ill-Posed Problems Series:

Composite Type Equations and Inverse Problems
A.I. Kozhanov

Inverse Problems of Vibrational Spectroscopy
A.G. Yagola, I.V. Kochikov, G.M. Kuramshina and Yu.A. Pentin

Elements of the Theory of Inverse Problems
A.M. Denisov

Volterra Equations and Inverse Problems
A.L. Bughgeim

Small Parameter Method in Multidimensional Inverse Problems
A.S. Barashkov

Regularization, Uniqueness and Existence of Volterra Equations of the First Kind
A. Asanov

Methods for Solution of Nonlinear Operator Equations
V.P. Tanana

Inverse and Ill-Posed Sources Problems
Yu.E. Anikonov, B.A. Bubnov and G.N. Erokhin

Methods for Solving Operator Equations
V.P. Tanana

Nonclassical and Inverse Problems for Pseudoparabolic Equations
A. Asanov and E.R. Atamanov

Formulas in Inverse and Ill-Posed Problems
Yu.E. Anikonov

Inverse Logarithmic Potential Problem
V.G. Cherednichenko

Multidimensional Inverse and Ill-Posed Problems for Differential Equations
Yu.E. Anikonov

Ill-Posed Problems with A Priori Information
V.V. Vasin and A.L. Ageev

Integral Geometry of Tensor Fields
V.A. Sharafutdinov

Inverse Problems for Maxwell's Equations
V.G. Romanov and S.I. Kabanikhin

Inverse and Ill-Posed Problems Series

Inverse Problems of Electromagnetic Geophysical Fields

P.S. Martyshko

///VSP///

Utrecht, The Netherlands, 1999

VSP BV
P.O. Box 346
3700 AH Zeist
The Netherlands

Tel: +31 30 692 5790
Fax: +31 30 693 2081
E-mail: vsppub@compuserve.com
Home Page: http://www.vsppub.com

First published in 1999

ISBN 90-6764-306-8

Contents

Introduction

Geophysical electromagnetic methods are widely used for studying the deep structure of the Earth, and for exploration and solution of ecological problems. The final stage of geophysical work is data processing. The data are obtained as a result of measurement or from application of a certain methodology.

One of the problems that arises in data processing is increasing the solution properties of the methods of interpretation. Development of effective mathematical theory of interpretation of geophysical fields and the theory of methods based on it may be useful.

Significant results in solving problems in the theory of electromagnetic geophysical fields have been obtained by Berdichevskii, Dmitriev, Zhdanov, Strakhov, Svetov, and others (see Bibliography).

The essential results are obtained in modeling electromagnetic fields (the direct problems). In comparison with the development of modeling methods, progress in solution of three-dimensional inverse problems is significantly less. This is so because we need new theoretical research and essential calculating resources for realization of algorithms of such problems. Recently, both Russian and other scientists have produced work in this field. The problems that arise in this case, from a mathematical point of view, are reduced to solution of inverse problems of mathematical physics. These problems are ill-posed; this fact essentially complicates anomaly fields processing and determination of the source parameters. The basis of this theory and the methods of solution of ill-posed problems were developed by Tikhonov (1943, 1963), Tikhonov and Glasko (1965), Tikhonov and Arsenin (1979), Ivanov (1956, 1962, 1963, 1967), and Lavrentiev (1956 a, b, 1960, 1962, 1965). The methods of regularization in application to geophysical problems were studied by Strakhov (1967, 1969, 1974), Glasko (1971, 1976, 1977), *et al.*

Note that, as follows from equivalence theory, see Tsirul'skii (1969), Strakhov (1973), Tsirul'skii and Nikonova (1975), Nadirashvili (1982), Cherednichenko (1983), in order to solve the inverse problem uniquely, i.e., to obtain the physical parameters of the object (density, magnetic permeability, and others), we need to use additional information. In this situation, the electromagnetic methods are the most favorable. These methods allow us to solve the inverse problem uniquely interpreting the observation data for some positions of the source, i.e. using only geophysical information.

Essential progress in research of theoretical problems and development of interpretation of the profile data of gravimagnetometry was reached by using the apparatus of complex functions theory, see Zamorev (1941), Ivanov (1956 a), Tsirul'skii (1963, 1969, 1974), Strakhov (1970, 1973), Tsirul'skii and Nikonova (1975), Strakhov and Blokh (1977 a, b), Nikonova and Tsirul'skii (1978). This allowed us to investigate the analytic properties of solution of direct two-dimensional problems of gravimagnetometry. Furthermore, certain classes of singular sources were defined. The inverse problem for the potentials of the basis of this theory were developed as two-stage methods of interpretation.

The essence of these two-stage methods is as follows. First, the observed data are approximated by the fields of singular sources (after an idea of Tsirul'skii and Strakhov). In the second stage we solve the theoretical inverse problem; i.e. given the field generated by singular sources, we construct the equivalent set of solutions. Thus, in the first stage, we divide the fields; then, the inverse problem is solved for each group of sources which are considered as one anomalous object. Tsirul'skii suggested using classes of singular sources for which the theoretical inverse problem is solvable: it may be reduced to solution of a finite system of equations relative to the parameters of the function $z(t)$. This function $z(t)$ maps conformally the knotty disk of the auxiliary complex plane t onto the desired domain. Note that the determination of $z(t)$ is equivalent to determination of the boundary of the anomaly domain. In this case, solvability of the inverse problem for the concrete class of potentials was established by means of the integral equation of Ivanov (1956 a). Note that the Strakhov method (which is a rather interesting method) is also a two-stage method.

Further, the use of the complex functions theory allowed Tsirul'skii to research certain analytic properties of solution of direct problems of the artificial magnetization method (MAM). Uniqueness of solution of the inverse problem of this method was begun by Lavrentiev (1956 b) and then by Atakhodzhaev (1966), and was also established by the complex function theory.

Namely, this theory allowed Strakhov to consider the problems of existence of equivalent sets of solutions of the inverse MAM problem without taking into account the demagnetization (Strakhov, 1977). It should be noted that the first algorithm of solution of the inverse two-dimensional MAM problem was also developed by using the complex functions theory (Strakhov and Blokh, 1977 b). Further, the author developed algorithms of solution of inverse MAM problems (taking account of demagnetization) which allowed construction of equivalent sets of solutions; the problems of solvability of the inverse problem in a finite form have also been investigated (Martyshko, 1981, 1982 a, b; Martyshko and Prutkin, 1982). Note that these results have universal character as MAM problems, the methods (Blokh, 1975, 1981, 1987; Blokh *et al.*, 1981) and electroprospecting for direct current sources are equivalent mathematically.

Unfortunately, the apparatus of complex function theory cannot be used effectively in the three-dimensional case: we have no analogs of the function $z(t)$, and we cannot select the classes of solvability in a finite form. However, for three-dimensional inverse gravimagnetometry problems, the two-stage methods of solution may be realized. Namely, we may select the sources in a certain class of singular sources and we may choose the effective parametrization of the boundary of the desired domain. As an equation of the inverse problem we use the representation of the potential by a surface integral along the boundary of the domain (see, Tsirul'skii and Prutkin, 1981; Tsirul'skii, 1990).

However, even the direct MAM problem, taking account of demagnetization (electric exploration by a direct current), i.e. the problem of determination of the sources W, external potential V_1 and internal potential V_2 of the domain D is reduced to the following boundary-value problem:

$$\left.\begin{array}{l} \Delta V_k = 0, \quad (k = 1, 2) \\ \\ V_1 = V_2, \\ \\ \mu_2 \dfrac{\partial V_2}{\partial n} - \mu_1 \dfrac{\partial V_1}{\partial n} = (\mu_1 - \mu_2) \dfrac{\partial W}{\partial n} \end{array}\right\} \text{ in } \partial D \qquad (0.1)$$

Here, D is the domain comprising material of constant magnetic permeability μ_2. The domain D is situated in the medium with a constant magnetic permeability μ_1. This problem, in the general case, has no solution in closed form, i.e. it is impossible to obtain the representation V_1 as an integral with the integrand given explicitly. The classic method of solution of direct problem (0.1) is that the solution is sought as a potential of simple layer or double layer. The unknown density of this potential is determined from an integral Fredholm equation of the second kind; the integral is taken along the body boundary (see, Filatov, 1969 b; Voskoboinikov, 1973; Mayergoyz, 1979; Martyshko, 1982 b). Another method is to use the equation relative to the magnetization of the whole volume (see, Vogel, 1963; Talvani, 1965; Sharma, 1966; Sobolev and Belogolov, 1968; Blokh, 1975, 1981, 1987). Further, we shall discuss the advantages of each of these methods.

Thus, we see that the inverse problem of electroprospecting, even for the electric field of a direct current, is reduced, in the general case, to the equation with the operator given implicitly; this holds all the more for variable electromagnetic fields.

So, we have the three main situations which are necessary for development of two-stage methods of solution of inverse problems. First, selection of the classes of solvability of the theoretical inverse problem in a finite form. Second, the explicit equations for the theoretical inverse problem. Third, representations of potentials by Cauchy-type integrals. Up to now we have realized only the third moment — we have the three dimensional analogs of the integrals of the Cauchy type (see, Strakhov, 1948; Vekua, 1959; Henl *et al.*, 1964; Yanovskii, 1978; Zhdanov, 1984). As, in the two-stage process, we have the anomaly field given everywhere, and for the electromagnetic fields we have obtained the analogs of the integrals of the Cauchy type; it may be possible to obtain explicit equations of the inverse problems and, therefore, to construct the algorithms for their solution.

Chapter 1.

Two-dimensional problems of a magnetic exploration method involving artificial field magnetization and electric exploration by a direct current

1.1. PROBLEM STATEMENT

We should recall that MAM problems, magnetic exploration involving demagnetization and electric exploration by a direct current are mathematically equivalent. Therefore, the results that emerge from research on MAM problems, have universal character (in the sense specified above). The two-dimensional MAM problems have been investigated by Filatov (1968, 1969, 1969 a), Tsirul'skii (1972, 1974 b, 1975), Strakhov (1977), Blokh (1975), Strakhov and Blokh (1977 a, b), and others. V. A. Filatov and Yu. I. Blokh developed the algorithm numerical solution of the direct problem. Research into broader questions connected with direct and inverse MAM problems taking account of demagnetization was carried out by Tsirul'skii (1972, 1974, 1975). Strakhov (1977) investigated questions of the existence and actual finding of the equivalent solutions of an inverse problem without taking account of demagnetization; thus, both authors used the technique of the theory of functions of a complex variable. With the CVFT technique, it is possible to investigate basic MAM problems more completely, to develop rather simple and effective methods for the solution of direct and inverse problems that, in turn, allows us to study a number of basic questions: the existence and actual finding of equivalent families of solutions of an inverse problem at one situation of a source, possibilities of practical application of the unique solution by use of two situations of a source.

Mathematically the direct problems are MAM formulated as follows. Suppose that area $D \in R^2$ with boundary ∂D, which is a regular analytical curve L (the class of such areas shall be designated through D_R) is filled by a substance with a constant magnetic permeability μ_2 and is in environment with a magnetic permeability μ_1. Let W be the magnetic potential of an inducing field, V_1 external and V_2 internal potentials of the area D, be induced in a field W. Then the finding of potentials V_1 and V_2 in statements 1 (taking account of demagnetization) and 2 (without account of demagnetization) is reduced to the solution of the following boundary problems, respectively:

$$\Delta V_k = 0, \quad (k = 1, 2)$$

$$\left. \begin{aligned} V_1 &= V_2, \\ \mu_2 \frac{\partial V_2}{\partial n} - \mu_1 \frac{\partial V_1}{\partial n} &= (\mu_1 - \mu_2) \frac{\partial W}{\partial n} \end{aligned} \right\} \tag{1.1}$$

$$\Delta V_k = 0, \quad (k = 1, 2)$$

$$\left. \begin{aligned} V_1 &= V_2, \\ \frac{\partial V_2}{\partial n} - \frac{\partial V_1}{\partial n} &= -4\pi æ \frac{\partial W}{\partial n} \end{aligned} \right\} \tag{1.2}$$

where æ is the magnetic susceptibility.

In research of two-dimensional problems, it is supposed that the inducing field is produced by an indefinitely long cable, the trace of which is in some point $a \in D^-$. The solution of the problem (1.2) is the potential of a simple layer with density $-æ\,(\partial W/\partial n)$, and the solution of problem (1.1) is reduced to the solution of some auxiliary Fredholm equation of the second kind (Frank and Mizes, 1937). This equation was examined by Filatov

(1969 b). He created an algorithm for numerical solution of the equation on the basis of a method of quickest release. The proposed method gives the numerical solution of the problem, but does not enable us to carry out research of the analytical properties of the required functions.

1.2. DIRECT PROBLEMS OF AN ELECTRIC EXPLORATION AND MAM AS PROBLEMS OF LINEAR CONJUNCTION OF ANALYTICAL FUNCTIONS

We shall move to complex coordinates $z = x + iy$, $\overline{z} = x - iy$, and consider functions of complex variable

$$u_1(z) = -\frac{1}{\pi}\left(\frac{\partial V_1}{\partial x} - i\frac{\partial V_1}{\partial y}\right)$$
$$u_2(z) = -\frac{1}{\pi}\left(\frac{\partial V_2}{\partial x} - i\frac{\partial V_2}{\partial y}\right) \tag{1.3}$$
$$u_3(z) = -\frac{1}{\pi}\left(\frac{\partial W}{\partial x} - i\frac{\partial W}{\partial y}\right).$$

Analyticity of functions $u_1(z)$ in D^-, $u_2(z)$ and $u_3(z)$ in D^+ follows from (1.3); moreover, $\lim\limits_{|z|\to\infty}\left(z^2 u_1(z)\right) = \text{const}$. Function $u_1(z)$ in (1.3) has been introduced for the first time in research of potential fields by Zamorev (1941); hereinafter, it was studied in research of problems in gravimagnetometry in the works of Ivanov (1956 a, b), Rapoport (1940), Strakhov (1970, 1973, 1977), Tsirul'skii (1963, 1969, 1972, 1974 a, b, 1975, 1976), Golizdra (1966), and others. It comprises the information on external potential $V_1(x, y)$ and can be investigated instead of it. We shall refer to this function as complex intensity of a magnetic field.

In Tsirul'skii and Prutkin (1981) it was shown that if $D \in D_R$, the problem (1.1) is reduced to the following problem of linear conjunction of analytical functions $u_1(z)$ and $u_2(z)$:

$$\lambda u_1(\zeta) + (\lambda + 1)u_2(\zeta) = -\overline{u_1(\zeta)}f'(\zeta) + u_3(\zeta) - \overline{u_3(\zeta)}f'(\zeta), \ \zeta \in L \tag{1.4}$$

where $f'(\zeta)$ is the derivative of the right-hand side of the equation of a curve L in complex coordinates,

$$\lambda = \frac{\mu_2 + \mu_1}{\mu_2 - \mu_1}. \tag{1.5}$$

The problem (1.4) is a special case of a general problem of linear conjunction, introduced by Markushevich (1946)

$$\varphi^+(\zeta) = a(\zeta)\varphi^-(\zeta) + b(\zeta)\overline{\varphi^-(\zeta)} + c(\zeta), \ \varphi^-(\infty) = 0 \tag{1.6}$$

at

$$a(\zeta) = \frac{\lambda}{\lambda+1}, \quad b(\zeta) = \frac{f'(\zeta)}{\lambda+1}, \quad c(\zeta) = \frac{\overline{u_3(\zeta)}f'(\zeta) - u_3(\zeta)}{\lambda+1}.$$

The problem was investigated by L. G. Mikhailov. In particular, he proved that at $|a(\zeta)| > |b(\zeta)|$ (elliptic case) under some restrictions on factors, the problem (1.6) is solvable and has a unique solution. As $|f'(\zeta)| = 1$, on L $\lambda = \text{const} > 1$ and $Ind\ a(\zeta) = 0$, since $a(\zeta) = \text{const}$, all conditions of Theorem 4.3 (Mikhailov, 1963) hold true, and the problem (1.4) has a unique solution. In (Tsirul'skii, 1974) it is further shown that the boundary problem (1.4) and a Fredholm equation for a simple layer are equivalent. The equation (1.4) allows us to carry out certain analysis of analytical properties of complex intensity $u_1(z)$, in particular, to consider the problem of peculiarities of analytical continuation $u_1(z)$ in D (Tsirul'skii, 1974).

When æ is small (i.e. in statement 2), problem (1.2) can be reduced to a boundary problem

$$-u_1(\zeta) + u_2(\zeta) = 2\pi æ \left[u_3(\zeta) - \overline{u_3(\zeta)}f'(\zeta)\right], \quad \zeta \in L. \tag{1.7}$$

1.3. EFFECTIVE ALGORITHM OF THE SOLUTION OF A PROBLEM OF LINEAR CONJUNCTION

The boundary problem (1.7) is a so-called problem about discontinuity of analytical functions: it is possible at once to perceive representation $u_1(z)$ as a Cauchy-type integral (Tsirul'skii, 1976):

$$u_1(z) = \frac{2\pi æ}{2\pi i} \int_L \frac{\overline{u_3(\zeta)}f'(\zeta)}{\zeta - z} d\zeta, \quad z \in D^-. \tag{1.8}$$

In the case when the external field is induced by a point source (Tsirul'skii, 1975),

$$u_3(z) = \frac{A \cdot i}{z - a}. \tag{1.9}$$

Substituting (1.9) in (1.8), we derive (Tsirul'skii, 1976)

$$u_1(z) = \frac{2\pi æ}{2\pi i} \int\limits_L \frac{A \cdot i \cdot f'(\zeta)}{(f(\zeta) - a)(\zeta - z)} \mathrm{d}\zeta. \tag{1.10}$$

Thus, (1.8) gives the solution of a direct problem in statement 2 in the closed form for any area $D \in D_R$. The solution of the boundary problem (1.4) for any $D \in D_R$ cannot be obtained in closed form. The effective numerical solution of problem (1.4) can be derived as follows. We shall designate

$$\frac{\lambda + 1}{\lambda} u_2(z) = u^+(z),\ u_1(z) = u^-(z),\ \frac{\overline{u_3} f' - u_3}{\lambda} \varphi(\zeta); \tag{1.11}$$

then from (1.4) we have

$$u^+(\zeta) = u^-(\zeta) + \overline{u^-(\zeta)} f'(\zeta)/\lambda + \varphi(\zeta). \tag{1.12}$$

We shall seek the solution (1.12) as a Cauchy-type integral

$$u(z) = \frac{1}{2\pi i} \int\limits_L \frac{\Psi(\zeta)}{\zeta - z} \mathrm{d}\zeta \tag{1.13}$$

where density of integral $\Psi(\zeta)$ is satisfied on a curve L with a Hölder condition with a parameter α, $0 < \alpha < 1$ (Chudinova, 1965). Designating through $A\Psi$ the singular operator

$$A\Psi = \frac{1}{\pi i} \int\limits_L \frac{\Psi(\zeta)}{\zeta - z} \mathrm{d}\zeta \tag{1.14}$$

and taking into account that on L (under the Sokhotskey – Plemel formulas) $u^+ = (\Psi + A\Psi)/2$, $u^- = (-\Psi + A\Psi)/2$, $\Psi = u^+ - u^-$, $A\Psi = u^+ + u^-$, we shall obtain from (1.12) the following singular equation for the required density of the Cauchy-type integral

$$\Psi(\zeta_0) = \frac{f'(\zeta_0)}{2\lambda} \overline{(-\Psi(\zeta_0) + A\Psi(\zeta))} + \varphi(\zeta_0) \tag{1.15}$$

which is equivalent to an edge(boundary) problem (1.12).

The solution of the equation (1.15) can be derived by the method of consecutive approximations

$$\Psi_n(\zeta_0) = \frac{f'(\zeta_0)}{2\lambda}\overline{(-\Psi_{n-1}(\zeta_0) + A\Psi_{n-1}(\zeta))} + \varphi(\zeta_0), \; \Psi_0 = \varphi. \tag{1.16}$$

We shall express (1.16) as $\Psi_n = B\Psi_{n-1} + \varphi$, $\Psi_0 = \varphi$. We have $\Psi_n = \sum\limits_{k=0}^{n} B^k\varphi$, if we accept $B^0 = E$, the identical operator. For required density Ψ we have

$$\Psi = \lim_{n\to\infty} \Psi_n = \sum_{k=0}^{\infty} B^k\varphi \tag{1.17}$$

if the limit exists.

We shall note that in (1.14) it is possible to avoid calculation of the integral with singularity in a point $\zeta_0 \in L$. For this purpose, it is sufficient that for function Ψ, satisfying a Hölder condition on the curve L, the formula

$$\frac{1}{\pi i}\int\limits_L \frac{\Psi(\zeta)}{\zeta - \zeta_0}\,\mathrm{d}\zeta = \frac{1}{\pi i}\int\limits_L \frac{\Psi(\zeta) - \Psi(\zeta_0)}{\zeta - \zeta_0}\,\mathrm{d}\zeta + \varphi(\zeta_0). \tag{1.18}$$

holds.

Then, it is possible to express (1.16) in the form

$$\Psi_n(\zeta_0) = \frac{f'(\zeta_0)}{2\lambda}\frac{1}{\pi i}\int\limits_L \overline{\frac{\Psi_{n-1}(\zeta) - \Psi_{n-1}(\zeta_0)}{\zeta - \zeta_0}}\,\mathrm{d}\zeta + \varphi(\zeta_0). \tag{1.19}$$

It is also possible to derive a sufficient condition of convergence of the method of consecutive approximations (1.19). We shall consider for this purpose Banach's space of functions, determined on the curves L and satisfying on it the Hölder condition with a parameter α, and with the norm

$$\| \Psi \|_\alpha = \max | \Psi(\zeta) | + \sup \frac{| \Psi(\zeta_2) - \Psi(\zeta_1) |}{| \zeta_2 - \zeta_1 |^\alpha}.$$

It is known (Muskhelishvili, 1962) that the estimate

$$\left| \frac{1}{\pi i} \int\limits_L \frac{\Psi_{n-1}(\zeta) - \Psi_{n-1}(\zeta_0)}{\zeta - \zeta_0} \mathrm{d}\zeta \right| \leq B_\alpha \parallel \Psi_{n-1} \parallel$$

holds and B_α is independent of Ψ. Then the following sequence of inequalities may be established:

$$\parallel \Psi_n - \Psi_{n-1} \parallel_\alpha = \left\| \frac{f'(\zeta_0)}{2\lambda} \frac{1}{\pi i} \int\limits_L \frac{(\Psi_{n-1}(\zeta) - \Psi_{n-2}(\zeta)) - (\Psi_{n-1}(\zeta_0)}{\zeta - \zeta_0} \right.$$

$$\left. - \frac{\Psi_{n-2}(\zeta_0))}{\zeta - \zeta_0} \mathrm{d}\zeta \right\|_\alpha \leq \frac{\parallel f' \parallel_\alpha B_\alpha}{2\pi\lambda} \parallel \Psi_{n-1} - \Psi_{n-2} \parallel_\alpha$$

$$\leq \ldots \leq \left(\frac{\parallel f' \parallel_\alpha B_\alpha}{2\pi\lambda} \right)^{n-1} \parallel \Psi_1 - \Psi_0 \parallel_\alpha$$

from which follows the sufficient condition of convergence

$$\frac{\parallel f' \parallel_\alpha B_\alpha}{2\pi\lambda} \prec 1.$$

The method of solution of a boundary problem (1.12) for areas of a class in a similar magnetic field follows, and in a field of a point source may be numerically realized. For the solution of a direct problem we consider (as in Rapoport, 1940; Ivanov, 1956 b; Strakhov, 1970, 1977; Tsirul'skii, 1976; Nikonova and Tsirul'skii, 1978) parametrization of the area D as $z = z(t)$, where $z(t)$ is a conform display of the disk $T : \{t : |t| < 1\}$ of an auxiliary complex plane t to area D, and $z(0) = 0$, $z'(0) > 0$ is known. Therefore, in (1.19), the integral on L can be replaced by an integral on the disk Γ using (after Vekua, 1959)

$$\frac{1}{\pi i} \int\limits_L \frac{\Psi(\zeta)}{\zeta - \zeta_0} \mathrm{d}\zeta = \frac{1}{\pi i} \int\limits_\Gamma \frac{\Psi(\zeta(t))\zeta'(t)}{\zeta(t) - \zeta_0} \mathrm{d}t \tag{1.20}$$

and for $f'(\zeta)$ we may use the formula (Himmelblau, 1975):

$$f'(\zeta(t)) = \overline{\zeta'(t)}/(t^2\zeta'(t)). \tag{1.21}$$

The algorithm (1.19), (1.13) taking account of (1.20), (1.21) is program realized, the program being written in FORTRAN language. The process continues until $\dfrac{\|\Psi_n - \Psi_{n-1}\|}{\|\Psi_n\|}$ becomes less than a certain ε.

We shall consider examples of algorithm application on theoretical models. First, we should note that for an ellipse in a similar stimulating field, the sum (1.17) can be calculated in a closed form, due to the analyticity of $f'(z)$ in D; that is, to state $u_1(z)$ explicitly

$$u_1(z) = \frac{\lambda \overline{H_0} + \overline{f'(\infty)} H_0}{\lambda - \overline{f'(\infty)^2}}[f'(\infty) - f'(z)]. \tag{1.22}$$

If the semiaxes of the ellipse are directed along the coordinate axes, then

$$f'(z) = \frac{a^2 + b^2}{c^2} \pm \frac{2ab}{c^2} \frac{z}{\sqrt{z^2 - c^2}}, \quad f'(\infty) = \frac{a-b}{a+b} \tag{1.23}$$

(here a and b are semiaxes of an ellipse, $c^2 = a^2 - b^2$); if, besides, H is real (the external field is directed along an axis of the ellipse), then

$$u_1(z) = \frac{H_0(\alpha - f'(z))}{\lambda - \alpha}, \quad \alpha = \frac{a-b}{a+b}. \tag{1.24}$$

From (1.24), for $\lambda = 1$, we get the established formula (Undezenkov, 1978) for the complex intensity of a perfectly conducting ellipse

$$u_1(z) = \frac{2H_0(\alpha + 1)R^2}{\sqrt{z^2 - 4\alpha R^2}(z + \sqrt{z^2 - 4\alpha R^2})}$$

$$R = \frac{1}{2}(a+b). \tag{1.24'}$$

An ellipse with centre in the origin ($a = 3$, $b = 1$) was taken as an example. Values were calculated by program realization of the circuit (1.19), (1.13) and values $u_1(z)$ from the formula (1.24) on a profile $z = x + i3$ were obtained. The maximum deviation fource was 10^{-5}, which shows high computing accuracy of the algorithm. It is known that complex intensity of an ellipse without account of demagnetization (Tsirul'skii, 1963) is as follows:

$$u_1(z) = 2\pi H_0[\alpha - f'(z)]. \tag{1.25}$$

From (1.24) and (1.25) we get

$$\frac{u(z)}{u_1(z)} = 1 + 2\pi æ - 2\pi æ\alpha. \tag{1.26}$$

As may be seen from (1.26), the influence of demagnetization is increased with increase of æ and reduction in α.

Let now the area D is transformed by conformal function of a disk $|t| < 1$ of an auxiliary complex plane t onto a plane z:

$$z(t) = B \ln\left(\frac{1 - \bar{d}t}{1 + \bar{\beta}t}\right). \tag{1.27}$$

In this case, complex intensity of area D without account of demagnetization is as follows:

$$u(z) = A_1\left(\frac{1}{z-a} - \frac{1}{z-b}\right)$$

and $A_1 = -2\pi æ B$, $a = z(d)$, $b = z(\alpha)$.

In Figure 1.1 are the resulting curve $V_1 = \Im(\pi u(z))$ areas (1.27) without taking account of (2) and including (1) the demagnetization effect. The curves I and II differ greatly already at æ $= 0.07$ (maximum divergence 12%).

The formula (1.26) and Figure 1.1 illustrate only quantitative distinction of results of the solution of the direct problem with and without account of the demagnetization effect. Actually, as may be seen from comparison of the formulas (1.22) and (1.24), even in the case of similar, but arbitrarily oriented external field, curve V_Y of an ellipse differs qualitatively. In Figure 1.2 we see that the curves V_Y of an ellipse with æ $= 0.1$ are brought into a similar external field H_0 without (2) and with account of (1). This distinction especially is essential for any area from class D_R in the field of a point source. Figure 1.3 shows curves of vertical intensity for the area, the function $z(t)$ resembling (1.27), without (2) and with (1) account of demagnetization. The situation of a source is marked by the letter a.

The described algorithm has been tested on a large number of models. The number of iterations in (1.19) is increased with growth of magnetic susceptibility æ and greater extent of a body. So, for an ellipse with a ratio of semiaxes 1:10 and æ $= 0.2$ it is necessary to carry out 6 iterations, while for an ellipse with ratio 1:3, only 2 iterations are required.

1.4. THE THEORETICAL INVERSE MAM PROBLEM AND ELECTRIC EXPLORATION WITH A DIRECT CURRENT

We shall define as a theoretical inverse MAM problem (electric exploration), in view of the demagnetization effect, the following problem: given a harmonic function $V_1(x,y)$ and connected with it by a relation (1.3) a function $u_1(z)$, analytical in the top half-plane and

$$\lim_{|z|\to\infty} z^2 u_1(z) = C_0 = \text{const}.$$

It is required to find an equivalent family of areas with a constant magnetic susceptibility, on boundaries of which $u_1(z)$ satisfies condition (1.4) (or (1.7)). We shall emphasize that this problem has great practical importance, as the classes of functions $u_1(z)$, for which a theoretical inverse problem (TIP) can be effectively solved, are convenient to use in the representation of an observed field (similar to those encountered in problems of gravimetry and magnetometry without taking account of demagnetization (Nikonova and Tsirul'skii, 1978)).

Let $D \in D_R$ be an area and the function $z(t)$ realize conformal display of a disk T with the boundary Γ on the area D.

We shall introduce the function $z^*(t) = \overline{z(1/t)}$, analytical in T^-; $z^*(t) = \overline{z(t)}$ in Γ. The function $z(t)$ can be considered as being subject to determination at the solution of the TIP, its finding implying actual and effective finding of the area D.

Chudinova (1965) derived the following integral equation of inverse problem 1 relative to $z(t)$:

$$\frac{\lambda}{2\pi i}\int\limits_{\Gamma} \frac{\tau z'(\tau) u_1(z(\tau))}{\tau - t}\,\mathrm{d}\tau = \frac{1}{2\pi i}\int\limits_{\Gamma} \frac{\overline{u_1(z(\tau)) z'(\tau)}}{\tau(\tau - t)}\,\mathrm{d}\tau - \frac{\overline{u_3(z(t)) z'(t)}}{t}. \tag{1.28}$$

If we can neglect the demagnetization effect, instead of (1.28) we have (see Tsirul'skii, 1976; Strakhov, 1977)

$$\frac{\lambda}{2\pi i}\int\limits_{\Gamma} \frac{u_1(z(\tau)) z'(\tau)}{\tau - t}\,\mathrm{d}\tau = 2\pi æ \left(\frac{\overline{u_3(z(t)) z'(t)}}{t^2}\right). \tag{1.29}$$

In the homogeneous external field $u_3 = H_0$ we come to the magnetic analogue of the integral equation of Ivanov (1956a). On the basis of the Ivanov equation, classes of complex intensities were defined, for which the TIP is solvable in a finite form. As was shown and illustrated in (Tsirul'skii, 1976; Nikonova and Tsirul'skii, 1978), such classes of intensities are convenient to use for approximation of an observed field, as the inverse problem can be solved for them rather effectively and quickly.

In the case of a gravitational field of the most general class, for which the TIP is completely solvable, we have the formula

$$u_1(z) = \sum_{k=1}^{n}\sum_{j=1}^{n_k} \frac{\beta_{jk}}{(z-b_k)^j} + \sum_{k=1}^{N} \gamma_k \ln(z-a_k) \tag{1.30}$$

and

$$\sum_{k=1}^{N} \gamma_k = 0. \tag{1.31}$$

In the case of a magnetic field $u_1(z)$ of form (1.30) we have

$$u_1(z) = \sum_{k=1}^{n}\sum_{j=1}^{n_k} \frac{\beta_{jk}}{(z-b_k)^{j+1}} + \sum_{k=1}^{N} \gamma_k \ln(z-a_k) \tag{1.32}$$

and for $\{\gamma_k\}_1^N$, condition (1.31) holds. We shall designate the class of intensities (1.32) as $A(z)$. For this class it is possible to make some progress on a method of solution of equation (1.28).

So, let the intensity of some area $D \in D_R$, filled by a substance with a constant magnetic susceptibility æ, be represented by (1.32). Substituting (1.32) in (1.28), we have

$$\sum_{k=1}^{n}\sum_{j=1}^{n_k} \frac{\beta_{jk}\lambda}{2\pi i} \int_L \frac{\tau z'(\tau)\,\mathrm{d}\tau}{(z(\tau)-b_k)^{j+1}(\tau-t)} + \sum_{k=1}^{N} \frac{\gamma_k}{2\pi i} \int_L \frac{\tau z'(\tau)\,\mathrm{d}\tau}{(z(\tau)-a_k)(\tau-t)}$$

$$= \sum_{k=1}^{n}\sum_{j=1}^{n_k} \frac{\beta_{jk}}{2\pi i} \int_L \frac{\overline{z'(\tau)}\,\mathrm{d}\tau}{\tau(z^*(\tau)-b_k)^{j+1}(\tau-t)} + \sum_{k=1}^{N} \frac{\gamma_k}{2\pi i} \int_L \frac{\overline{z'(\tau)}\,\mathrm{d}\tau}{\tau(z^*(\tau)-\overline{a_k})(\tau-t)} \tag{1.33}$$

$$-\frac{\overline{u_3(z(t))z'(t)}}{t}.$$

For calculation of integrals from the left-hand side (1.33) we shall note that integrand functions are analytical at $|t| < 1$, except for the points c_k and α_k, points being prototypes b_k and a_k, accordingly, where they have poles of finite order. It follows from the relations

$$z(\tau) - b_k = (t - c_k)\Psi(\tau)$$

$$z(\tau) - a_k = (t - \alpha_k)\varphi(\tau)$$

where $\Psi(\tau)$ and $\varphi(\tau)$ are analytical functions, which have no zeros in the unit disk T by virtue of univalence of the function $z(t)$.

Using the formula for calculation of such integrals (Gakhov, 1963)

$$\Phi^-(t) = -\sum_k g_k\left(\frac{1}{t - c_k}\right)$$

where g_k is a polynomial representing the main part of the decomposition of the density of the Cauchy-type integral in the neighborhood of a final pole, we get

$$\sum_{k=1}^{n}\sum_{j=1}^{n_k}\frac{\beta_{jk}}{2\pi i}\int_L \frac{\tau z'(\tau)\,\mathrm{d}\tau}{(z(\tau) - b_k)^{j+1}(\tau - t)} = \sum_{k=1}^{n}\sum_{j=1}^{n_k}\frac{\beta_{jk}}{(t - c_k)^j} \tag{1.34}$$

$$\sum_{k=1}^{N}\frac{\gamma_k}{2\pi i}\int_L \frac{\tau z'(\tau)\,\mathrm{d}\tau}{(z(\tau) - a_k)(\tau - t)} = \sum_{k=1}^{N}\frac{\gamma_k \alpha_k}{\alpha_k - t} \tag{1.35}$$

and from the method of calculation it is clear that $z(c_k) = b_k$, $z(\alpha_k) = a_k$, $|c_k| < 1$, $|\alpha_k| < 1$.

For calculation of integrals in the right-hand side of (1.33) we shall note that the integrand function in the first of them is analytical in D^-, except for the points $t = 1/\overline{c_k}$, in which it has poles of finite order (as $z^*(1/\overline{c_k}) = \overline{b_k}$); the integrand function in the second integral is analytical in D, except for the points $t = 1/\overline{\alpha_k}$, where it has poles of the first order (as $z^*(1/\overline{\alpha_k}) = \overline{a_k}$).

Applying the formula for calculation of such integrals, we obtain

$$\Phi^-(t) = -\eta(t) + \sum_k g_k\left(\frac{1}{t - a_k}\right) + g(t)$$

where $\eta(t)$ is the density of the Cauchy-type integral, $g(t)$ is the body of decomposition η in a neighborhood of an infinitely removed point (in this case $g = 0$), we receive

$$\sum_{k=1}^{n}\sum_{j=1}^{n_k}\frac{\overline{\beta_{jk}}}{2\pi i}\int_L \frac{z'(\tau)\,\mathrm{d}\tau}{\tau(z^*(\tau)-\overline{b_k})^{j+1}(\tau - t)}$$

$$= \sum_{k=1}^{n}\sum_{j=1}^{n_k}\frac{\beta_{jk}\overline{z'(\tau)}}{t(z^*(t)-\overline{b_k})^{j+1}} + \sum_{k=1}^{n}\sum_{j=1}^{n_k+1}\frac{\theta_{jk}}{(t-1/\overline{c_k})^j} \tag{1.36}$$

$$\sum_{k=1}^{N}\overline{\gamma_k}\int_\Gamma \frac{\overline{z'(\tau)}\,\mathrm{d}\tau}{\tau(z^*(\tau)-\overline{a_k})(\tau - t)}$$

$$= -\sum_{k=1}^{N}\overline{\gamma_k}\frac{\overline{z'(\tau)}}{t(z^*(t)-a_k)} - \sum_{k=1}^{N}\frac{\overline{\gamma_k}/\overline{\alpha_k}}{t-1/\overline{\alpha_k}}. \tag{1.37}$$

Substituting (1.34)–(1.37) in (1.33), we get

$$\sum_{k=1}^{n}\sum_{j=1}^{n_k+1}\frac{p_{jk}\lambda}{(t-1/\overline{c_k})^j} + \sum_{k=1}^{N}\frac{\lambda\gamma_k\alpha_k}{\alpha_k - t} = -\sum_{k=1}^{n}\sum_{j=1}^{n_k}\frac{\overline{\beta_{jk}z'(\tau)}}{t(z^*(t)-\overline{b_k})^{j+1}}$$

$$+\sum_{k=1}^{n}\sum_{j=1}^{n_k+1}\frac{\theta_{jk}}{(t-1/\overline{c_k})^j} - \sum_{k=1}^{N}\overline{\gamma_k}\frac{\overline{z'(\tau)}}{t(z^*(t)-a_k)} + \sum_{k=1}^{N}\frac{\overline{\gamma_k}/\overline{\alpha_k}}{t-1/\overline{\alpha_k}} \tag{1.38}$$

$$-\frac{\overline{u_3(z(t))z'(t)}}{t}.$$

Putting $|t| = 1$ and applying operation of conjunction to both sides of (1.38), we get

$$\sum_{k=1}^{n}\sum_{j=1}^{n_k+1}(-1)^j\frac{\lambda p_{jk}t^j}{\overline{c_k}(t-1/\overline{c_k})^j} + \sum_{k=1}^{N}\frac{t\overline{\gamma_k}\lambda}{t-1/\overline{\alpha_k}}$$

$$= -\sum_{k=1}^{n}\sum_{j=1}^{n_k}\frac{\beta_{jk}t(dz/dt)}{(z(t)-b_k)^{j+1}} + \sum_{k=1}^{n}\sum_{j=1}^{n_k+1}(-1)^j\frac{t^j c_k^j\overline{\theta_{jk}}}{(t-1/c_k)^j} \tag{1.39}$$

$$-\sum_{k=1}^{N}\frac{\gamma_k(dz/dt)}{z(t)-a_k}+\sum_{k=1}^{N}\frac{\gamma_k t}{t-\alpha_k}-tu_3(z(t))\frac{\mathrm{d}z}{\mathrm{d}t},$$

a the differential equation with divided variables. After simple transformations from (1.39) we obtain

$$\left(\sum_{k=1}^{n}\sum_{j=1}^{n_k}\frac{\beta_{jk}}{(z-b_k)^{j+1}}+\sum_{k=1}^{N}\frac{\gamma_k}{z-a_k}+u_3(z)\right)\mathrm{d}z=\left(-\sum_{k=1}^{N}\frac{\lambda\overline{\gamma_k}}{t-1/\overline{\alpha_k}}\right.$$

$$\left.+\sum_{k=1}^{N}\frac{\gamma_k}{t-\alpha_k}+\sum_{k=1}^{n}\sum_{j=1}^{n_k+1}\frac{\lambda\varphi_{jk}}{(t-1/\overline{c_k})^j}+\sum_{k=1}^{n}\sum_{j=1}^{n_k}\frac{\Psi_{jk}}{(t-c_k)^j}\right)\mathrm{d}t \qquad (1.40)$$

where the coefficients φ_{kj} are expressed through b_k, β_k, c_k. Carrying out integration in (1.40), we have

$$\sum_{k=1}^{N}\gamma_k\ln(z-a_k)+F(z)-\sum_{k=1}^{n}\sum_{j=1}^{n_k}\frac{\beta_{jk}}{j(z-b_k)^j}+C$$

$$=\sum_{k=1}^{N}\gamma_k\ln(t-\alpha_k)-\lambda\sum_{k=1}^{N}\overline{\gamma_k}\ln(t-1/\overline{\alpha_k})+\lambda\sum_{k=1}^{N}\varphi_{k1}\ln(t-1/\overline{c_k}) \qquad (1.41)$$

$$-\lambda\sum_{k=1}^{n}\sum_{j=1}^{n_k+1}\frac{\varphi_{jk}}{(t-1/\overline{c_k})^{j-1}(j-1)}+\sum_{k=1}^{N}\Psi_{k1}\ln(t-c_k).$$

Here $F(z)$ is the primitive of the function $u_3(z)$, and the integration constant C is found by substitution of condition $z(0)=0$ in (1.41). Thus, the solution of the problem, the function $z(t)$, can be found from the transcendental equation (1.41). This equation brings in a number $N+n$ of unknown parameters α_k and c_k. For their determination we have $N+n$ additional conditions:

$$z(c_k)=b_k \quad (k=1,2,\dots,n)$$

$$z(\alpha_k)=a_k \quad (k=1,2,\dots,N). \qquad (1.42)$$

Thus, we have proved the following.

Theorem 1.1. *Let D be an area of the class D_R filled by a substance with a constant magnetic susceptibility æ. If its complex intensity $u_1(z)$, induced in the field $u_3(z)$, may be represented by (1.32), the solution of the problem, the function $z(t)$, satisfies the system of equations (1.41), (1.42).*

Remark 1.1. In this case, it is impossible to prove that the TIP is solvable in a finite way, that is, to establish an explicit representation of the function $z(t)$.

Remark 1.2. Chudinova (Tsirul'skii, 1975) had proved that the theorem was proved: if $u_3 = \text{const}$ and $u_1(z) = \sum_{k=1}^{n-1} c_k/z^{k+1}$, then $z(t)$ is an algebraic function. From the equation (1.41) as a special case, a result, generalizing this theorem follows: if $u_1(z)$ is a rational function, then $z(t)$ is algebraic.

Remark 1.3. Generally speaking, an equivalent family of solutions is possible to obtain by solving system (1.41), (1.42) for various values of æ. We should recall that parameter æ connects the magnetic permeabilities of containing environment μ_1 and magnetic object μ_2 by formula (1.5). We shall show now that the inverse MAM problem for class (1.32) in statement 2 has a finite solution. Substituting (1.32) in (1.29), we have:

$$\sum_{k=1}^{n}\sum_{j=1}^{n_k}\frac{\beta_{jk}}{2\pi i}\int_{\Gamma}\frac{z'(\tau)\,d\tau}{(z(\tau)-b_k)^{j+1}(\tau-t)}+\sum_{k=1}^{N}\frac{\gamma_k}{2\pi i}\int_{\Gamma}\frac{z'(\tau)\,d\tau}{(z(\tau)-a_k)(\tau-t)}$$
$$=-2\pi æ\frac{u_3(z)z'(t)}{t^2}. \tag{1.43}$$

Substituting expression (1.34), (1.35) in (1.43), we get

$$\sum_{k=1}^{n}\sum_{j=1}^{n_k+1}\frac{p_{jk}}{(t-1/c_k)^j}+\sum_{k=1}^{N}\frac{\gamma_k\alpha_k}{\alpha_k-t}=2\pi æ\overline{\frac{u_3(z)z'(t)}{t^2}} \tag{1.44}$$

and $z(t)$ satisfies system (1.42). Repeating the argument, with the help of which relation (1.42) was obtained, we have:

$$\sum_{k=1}^{n}\sum_{j=1}^{n_k}\frac{\varphi_{jk}}{(t-1/\overline{c_k})^j}+\sum_{k=1}^{N}\overline{\gamma_k}\ln(t-1/\overline{\alpha_k})=-2\pi æF(z) \tag{1.45}$$

where the coefficients φ_{jk} are expressed by p_{jk} and c_k, and $F(z)$ and C have the same sense as in (1.41). We shall designate the left part of equality (1.45), divided by $-2\pi æ$, by $R(t)$. For the complex intensity $u_1(z) \in A(z)$ we have established an explicit form of the function $z(t)$ for the required area D:

$$z(t) = F^{-1}(R(t)) \tag{1.46}$$

where $F^{-1}(z)$ is the function, inverse for $u_3(z)$; $z(t)$ satisfies system (1.42). We should recall, that $u_3 = Ai/(z-a)$. Then, according to (1.46), we have

$$z(t) = \exp\left(K(t)\right) + a \tag{1.47}$$

where

$$K(t) = \sum_{k=1}^{n}\sum_{j=1}^{n_k} \frac{\Psi_{jk}}{(t-1/\overline{c_k})^j} + \sum_{k=1}^{N} g_k \ln(t - 1/\overline{\alpha_k}) + C$$

and

$$\Psi_{jk} = \varphi_{jk}/(-2\pi æ Ai), \ g_k = \overline{\gamma_k}/(-2\pi æ Ai) \quad .$$

$$z(c_k) = b_k, \ z(\alpha_k) = a_k, \ \mid c_k \mid < 1, \ \sum_{k=1}^{N} g_k = 0. \tag{1.48}$$

Strakhov (1977) investigated the solvability of the inverse problem 2 in a class of areas, produced by conformal functions of the disk with the help of rational fractions

$$z(t) = A\frac{\prod\limits_{k=1}^{N}(t-\alpha_k)}{\prod\limits_{r=1}^{P}(t-c_r)} \tag{1.47$'$}.$$

Obviously, (1.47′) is a special case of (1.47). We shall prove now the converse statement. Let for the area D the function $z(t)$ be of form (1.47). From the results of Tsirul'skii (1974) and Strakhov (1977) it follows that the complex intensity of area D is calculated in this case by the formula

$$u(\zeta) = 2I æi \int\limits_{\Gamma} \frac{dR^*(t)}{dt} \frac{dt}{z(t) - \zeta} \tag{1.49}$$

where I is the force of the current in the cable. The function $R^*(t) = \overline{R(1/\bar{t})}$ has the form

$$R^*(t) = \sum_{k=1}^{n} \sum_{j=1}^{n_k} \frac{\mu_{jk}}{(t - 1/c_k)^{j+1}} + \sum_{k=1}^{N} \frac{g_k}{t - \alpha_k}. \tag{1.50}$$

We shall designate by $t(z)$ the function, inverse to $z(t)$. From the univalence of the function $z(t)$, univalence of $t(z)$ follows; therefore,

$$t(z) - c_k = (z - b_k)\varphi_k(z), \; t(z) - \alpha_k = (z - a_k)\Psi_k(z)$$

and the functions $\varphi_k(z)$ and $\Psi_k(z)$ are analytical in D and do not therein equal zero. We shall make under the integral in (1.49) a change of variable $t = t(z)$. Taking (1.50) into account, we obtain

$$u_1(z) = 2I æi \int\limits_{L} \left(\sum_{k=1}^{n} \sum_{j=1}^{n_k} \frac{(-j)\mu_{jk}}{(z - b_k)^{j+1}\varphi^{j+1}} + \sum_{k=1}^{N} \frac{\overline{g_k}}{(z - a_k)\Psi_k(z)} \right) \frac{t'(z)dz}{\zeta - z}.$$

After calculation of the integral we shall have

$$u_1(z) = \sum_{k=1}^{n} \sum_{j=1}^{n_k} \frac{\beta_{jk}}{(z - b_k)^{j+1}} + \sum_{k=1}^{N} \frac{g_k}{z - a_k}.$$

Thus, we have proved the following

Theorem 1.2. *In order that $u_1(z) \in A(z)$, it is necessary and sufficient that $z(t)$ has a form (1.47).*

From this theorem it follows that the inverse MAM problem in statement 2 has a finite solution. The function $z(t)$, as follows from (1.47), is determined by $2N + n + m + 1$ parameters ($m = \sum_{k=1}^{N} n_k$), and C may be found from the condition $z(0) = 0$, $g_k = \overline{\gamma_k}/(-2\pi æ Ai)$, $k = \overline{1, N}$; from relation (1.42) we get $N + n$ of the equations. The deficient m equations can be as follows: we substitute $u_1(z)$ and $z(t)$ of the form (1.47) in the integral equation (1.29); equating factors at identical degrees $1/(t - c_k)$ in the left and right parts, we get the required m equations. To illustrate this, we shall consider a special case. Let

$$u_1(z) = \sum_{k=1}^{N} A_k \left(\frac{1}{z - a_k} - \frac{1}{z - b_k} \right). \tag{1.51}$$

Then

$$z(t) = \exp R(t) + a$$

where

$$R(t) = \sum_{k=1}^{N} g_k \left[\ln(t - 1/\overline{\alpha_k}) - \ln(t - 1/\overline{\beta_k}) \right] + C.$$

The coefficients $g_k = A_k/(-2\pi æ Ai)$, α_k and β_k are obtained from conditions $z(\alpha_k) = a_k$, $z(\beta_k) = b_k$. If g_k is a real number, then

$$z(t) = C_1 \prod_{k=1}^{N} \left(\frac{t - 1/\overline{\alpha_k}}{t - 1/\overline{\beta_k}} \right)^{g_k} + a. \tag{1.52}$$

If we consider magnetic susceptibility as a parameter, the relations obtained set the equivalent family of solutions of the inverse MAM problem without taking account of the demagnetization. We shall consider one example of such a family. In (1.51), let $N = 1$; then from (1.52) we have

$$z(t) = a \left[1 - \left(\frac{t - 1/\overline{\alpha_k}}{t - 1/\overline{\beta_k}} \right)^{g_k} \right]$$

$$z(\alpha_1) = a_1, \; z(\beta_1) = b_1, \; g_1 = A_1/(-2\pi æ Ai). \tag{1.53}$$

Without loss of generality, let us assume $a_1 = 0$; then $\alpha_1 = 0$, and to find β_1, we have the equation

$$a \left[1 - (1 - \beta_1^2)^{-g_1} \right] = b_1$$

whence

$$\beta_1 = \sqrt{1 - (a/(a-b_1))^{1/g_1}}$$

$$z(t) = a\left[1 - (1-\beta_1 t)^{-g_1}\right]. \tag{1.54}$$

In that specific case at $g_1 = -1$, $z(t) = a\beta_1 t$; that is, we have a disk with the radius $R = a\beta_1$. Thus, relations (1.53) set continuous with respect to æ a family of areas, equivalent to a disk; and in this family, only one area is a disk.

From the results discussed above it follows that we may develop a method of solution of the inverse MAM problem 2, based on finite solution of a TIP (similar to the method of solution of inverse problems of gravimetric and magnetic exploration in an earth field without account of demagnetization (Tsirul'skii, 1976; Nikonova and Tsirul'skii, 1978)). However, as is shown in Blokh (1975); Strakhov and Blokh (1977) and illustrated above, the influence of demagnetization has an effect on the solution already for a direct problem, and, thence, it should have an effect on the inverse problem solution. Therefore, from the practical point of view, it is more important to develop methods of solution for problem 1, though the result of Theorem 1.2 will be used for a choice of initial approximation for finding the solution of the inverse problem 1.

1.5. EFFECTIVE ALGORITHM FOR SOLVING THE INVERSE MAM PROBLEM TAKING ACCOUNT OF DEMAGNETIZATION

As was noted above (Remark 1.3), the solution of inverse problem 1 may be obtained by solution of system (1.41), (1.42) at various values of a parameter. At the same time, we could not develop an effective numerical method for solving this system. The difficulties were connected with indefinition arising on substitution in (1.41) of the values $t = \alpha_k$, $t = c_k$, which it is necessary to determine. On the other hand, the high speed of solution of the direct problem, and also the opportunity of good parametrization of the required area — description of boundary ∂D by the function $z(t)$ — have allowed us to undertake an attempt to solve the inverse problem on the basis of the method of solution of the direct problem elaborated above and one of algorithms of nonlinear programming. When solving the inverse problem, we shall seek $z(t)$ as

$$z(t) = \sum_{i=1}^{n} B_1 \ln\left[(1-\overline{\alpha_i}t)/(1-\overline{\beta_i}t)\right]. \tag{1.55}$$

This enables us to describe a rather wide set of areas from D_R by the minimum number of parameters. As was shown in (Nikonova and Tsirul'skii, 1978), the format of (1.51) has appeared convenient for selection of the derivative of magnetic potential. The value to use at the first stage of interpretation, the stage of selection of an observed field, the programs of (1.51) available, we shall illustrate by solving the inverse problem for $u_1(z)$ in form (1.51). Solution of the inverse problem is reduced to minimization of the functional

$$f(\mathbf{x}) = \sum_{i=1}^{N_1} |u_1(z_i) - u(z_i, \mathbf{x})|^2 \tag{1.56}$$

where $\{z_i\}_{i=1}^{N}$ are the points of the structure; $\mathbf{x} = \{B_j, \alpha_J, \beta_J\}$ is a vector of parameters of the required function $z(t)$ in the form (1.55) on which the target functional f is minimized; $u(z_i)$ is the right part of the equation (1.55). The functional f is minimized until the inequality

$$f(\mathbf{x}) < \varepsilon$$

begins to hold. Here $\varepsilon > 0$ is a certain number which will be carried over further. If the required accuracy is not achieved, in sum (1.56) a new term is added and the process of minimization is repeated. We begin the construction of the equivalent family from small values of the parameter æ. As zero approximation we take $z_0(t)$ of form (1.55), and α_i and β_i are from conditions

$$\begin{aligned} \widetilde{z}(\alpha_i) &= a_1, \\ \widetilde{z}(\beta_i) &= b_1. \end{aligned}$$

Here $\widetilde{z}(t)$ is the solution of the inverse problem 2, $B_i = \overline{A_i}/2\pi æ H_0$, if $u_3 = H_0$, and $B_i = a\overline{A_i}/(-2\pi æ Ai)$, if $u_i = Ai/(z-a)$. For the choice of initial approximation we have used Theorem 1.2, whence we obtain $z(t)$, and the following relation (small æ)

$$a\left[1-\left(\frac{1-\overline{\alpha_i}t}{1-\beta_i t}\right)^{g_i}\right]=a\left[1-\exp\left\{g_i\ln\left(\frac{1-\overline{\alpha_i}t}{1-\beta_i t}\right)\right\}\right]\approx$$
$$\approx a\left[-g_i\ln\left(\frac{1-\overline{\alpha_i}t}{1-\beta_i t}\right)\right]$$
$$\left(g_i=\overline{A_i}/(-2\pi æ Ai)\right).$$

Further, for construction of the equivalent family of solutions of the inverse problem on each step (of æ), for initial approximation we take the previous term of the family. We shall consider examples of realization of the algorithm. In Figure 1.4 we see the solutions of the inverse problem in an earth field for $u_1(z)$ of kind (1.51). They are constructed at $N=2$ for the value $æ=0.5$; the figures designate the solutions of the problem without taking account (2) and with account (1) of demagnetization. In Figure 1.5 the solutions of the inverse MAM problem in view of demagnetization for a number(line) of values æ are constructed at $N=2$ in formula (1.51). We see that the geometry of area with reduction æ (approximation of a source a to area) is deformed. In contrast to magnetic exploration, where at small æ the area becomes close to a disk (Ivanov, 1956 b), in MAM this does not occur. As the calculations show, if æ diminishes and the area becomes so large that it may embrace a source, there is infringement of the univalence of $z(t)$, and the curve L becomes a curve with self-crossing. The source appears inside a piece of boundary. In the examples mentioned above it was impossible to estimate the quality of work of the algorithm of the solution of the inverse problem, as the exact solution of inverse problem 1 for $u_1(z)$ of kind (1.51) is not known. In order to check the accuracy of the algorithm, the inverse problem of a magnetic exploration for $u_1(z)$ of kind (1.24) was solved (in this case the exact solution is an ellipse) and the result was compared with the exact solution with the obtained one. The results were satisfactory (Figure 1.6, 1-exact solution; 2-approached). In order to estimate the accuracy of the two-stage algorithm for interpretation of MAM data the following was done: the direct problem for an ellipse with $æ=0,1$, semiaxis $a=3$, $b=1$ was solved, the field calculated was approximated by the function $u_1(z)$ of a form (1.51) and the inverse problem was solved. The result was quite satisfactory (Figure 1.7, 1-exact solution, 2-approached). We shall consider now the special case $\lambda=1$ ($æ=\infty$) of problem 1; it is of interest to us as the limit case; thus, we obtain minimum (at $æ>0$) areas. For the solution of problems of an electric exploration, this case corresponds to an ideal conductor. The solution of direct and inverse problems in this case is simpler. As was shown by Tsirul'skii (1975), the solution of the direct problem may be obtained in the closed form for any area $D \in D_R$. Let the function $z_1(t)$ map conformally T^- on D^-. It is known that

$$z_1(t) = Rt + \Psi^-(t) \tag{1.57}$$

where $\Psi^-(t)$ is analytical in D^- and $R > 0$. Then (Tsirul'skii, 1969)

$$u_1(z) = \frac{\Psi_1(t(z)) - \Psi_2(t(z))}{t(z)z'(t(z))}$$

where

$$\Psi_1(t) = \frac{1}{2\pi i} \int_\Gamma \frac{\tau u_3(z(t))z'(\tau)}{\tau - t} \, d\tau$$

$$\Psi_2(t) = \frac{1}{2\pi i} \int_\Gamma \frac{u_3(z(t))z'(\tau)}{\tau - t} \, d\tau.$$

For $u_3(z)$ of form (1.9) and $u_3 = H_0$, the functions Ψ_1 and Ψ_2 may be calculated in an explicit form. In the second case, we have

$$u_1(z) = H_0(-1 + Rt'(z) + Rt'(z)/t^2(z)). \tag{1.58}$$

Equation (1.58) may be used for the solution of the theoretical inverse problem as follows. We shall designate by $F_1(z)$ the primitive of the function $u_1(z)$ $(F_1(\infty) = 0)$. Then

$$F_1(z) = H_0(-z_1(t) + Rt(z) - R/t(z)). \tag{1.59}$$

Making in (1.59) the change $z_1 = z_1(t)$, we have

$$F_1(z_1(t)) = H_0(-z_1(t) + R(t) - R/t). \tag{1.60}$$

Due to formula (1.57), we get

$$F_1(z_1(t)) = H_0(-\Psi^-(t) - R/t). \tag{1.61}$$

If we have $u_1(z)$ (and, therefore, $F_1(z)$) in explicit form, for example, as a result of an approximation, it is possible to use equation (1.61) for the solution of an inverse problem, that is, to find parameters of the function $z_1(t)$. In this case, we take

$$\Psi^-(t) = P_{-n}(t)$$

where

$$P_{-n}(t) = \sum_{i=1}^{n} a_i t^{-i}. \tag{1.62}$$

The solution of the inverse problem is reduced to minimization of the functional

$$f(\mathbf{x}) = \sum_{i=1}^{N} |F_1(z_1(t_i, \mathbf{x})) - \Pi(t_i, \mathbf{x})|^2 \tag{1.63}$$

of the parameters $\mathbf{x} = \{R, a_1, a_2, ..., a_n\}$ of the required function $z_1(t)$, where $\{t_i\}$ are the knots of the disk Γ, Π is the right-hand side of the equation (1.61). We shall notice that the choice of $z_1(t)$ in form (1.57), where $\Psi^-(t)$ has form (1.62), allows us even for the description of rather extended objects, to use $P_{-n}(t)$ of a minimal degree. So, for the description of the boundary of an ellipse (with any ratio of semiaxes!) it is enough to put $n = 1$:

$$z_1(t) = R(t + \alpha/t) = Rt + a_1/t \tag{1.64}$$

where R and α are connected with the semiaxes of the ellipse by formulas (1.24), (1.24′). This is so because on Γ the relation $1/t = \bar{t}$ is carried out and (1.64) is possible to rewrite as

$$z_1(t) = Rt + a_1 t = R\,(Re(t) + i \cdot Im(t)) + a_1\,(Re(t) - Im(t)) =$$
$$= (R + a_1)\,Re(t) + i\,(R - a_1)\,Im(t).$$

By the appropriate choice of R and a_1, we may achieve any ratio between the real and imaginary parts in the last sum, that is, to represent any extended object. We have realized the process of minimization of a functional by the generalized modified Gauss – Newton method (Ortega and Reinbold, 1975; Starostenko and Oganesyan, 1977). As testing of the algorithm has shown on a number of examples in (1.62), it is enough to put $n = 8$, in order to describe boundaries of a rather wide class of areas. For comparison, we shall note that it takes 5 times more program realization time for solution of the above algorithm in the solution of problem 1 to calculate the area with fixed æ. The function $z_1(t)$ in a form (1.64) is taken as an initial approximation for solving the inverse problem. Thus, R and a_1 are found from the conditions of equality of the first two coefficients in decomposition by degrees of $1/z$ in the right and left parts of the equation (1.58). In Fig. 1.8, the form of solution of the inverse problem for $u_1(z)$ in (1.51) is constructed, $N = 2$, $A_1 = 2$, $A_2 = 3$, $a_1 = 1,8 + 2i$, $b_1 = 4 + 0,6i$, $a_2 = -4,8 - 1,4i$, $b_2 = 1,2 - i$ at $\lambda = 1$.

1.6. UNIQUENESS OF THE SOLUTION OF THE INVERSE PROBLEM OF MAGNETIC EXPLORATION

From the results of Lavrentiev (1956 b) it follows that for the solution of the inverse problem of magnetic exploration (or electric exploration), which is to find a magnetic susceptibility æ (or conductivity σ) we need to carry out measurements of a secondary field at a continuously moving source of a primary field (this means that we should have repeated measurements). In a two-dimensional case with use of the CVFT apparatus it is possible to obtain more reliable results. Atakhodzhaev (1966) proved that, in a class of stellar areas, the stipulation of two gravitational potentials with density $\sigma = \text{const}$ and $\sigma = \sigma(x^2 + y^2)$ provides unique solution of the problem (finding area and size). The same theorem was proved by Tsirul'skii (1972) for areas of a class D_R. Filatov (1969 a) proved the theorem of uniqueness in a class of areas, close to a disk, for a case of two positions of the cable in the linearized statement. It is necessary to note that the disks are not equivalent families in the case of stipulation of a secondary field in one situation of a source. As was noted above, each of disks enters into some equivalent family, where all other areas are not disks. Let D_R be located in the lower half-plane, the point $a \in D^-$ and lies on a real axis. We shall assume that the function $z(t)$, realizing conformal display of a disk T to area D^- is analytical in some area $T' \supset T$ and that there exists a point $t \in T' \backslash T$ such that $z_0(t) = a$ $(z'(t_0) \neq 0)$. We shall define the class of such areas as $D_R(a)$ (Tsirul'skii, 1972). The following theorems were proved by A. V. Tsirul'skii.

Theorem 1.3. *Let D_1 and D_2 be areas of class $D_R(a)$ with boundaries L_1 and L_2, whose equations are $\overline{z} = f_1(z)$ and $\overline{z} = f_2(z)$. If everywhere outside of $D_1 \cup D_2$ the relations*

$$\frac{1}{2\pi i}\int\limits_{L_1} \frac{æ_1 f_1(\zeta)}{\zeta - z}\,\mathrm{d}\zeta = \frac{1}{2\pi i}\int\limits_{L_2} \frac{æ_2 f_2(\zeta)}{\zeta - z}\,\mathrm{d}\zeta$$

$$\frac{1}{2\pi i}\int\limits_{L_1} \frac{æ_1 \ln(f_1(\zeta) - a)}{\zeta - z}\,\mathrm{d}\zeta = \frac{1}{2\pi i}\int\limits_{L_2} \frac{æ_2 \ln(f_2(\zeta) - a)}{\zeta - z}\,\mathrm{d}\zeta$$

hold, where $æ_1 \geq æ_2 > 0$, then $æ_1 = æ_2$.

Theorem 1.4. *If under the same conditions as in Theorem 1.3 everywhere outside of $D_1 \cup D_2$ the relations*

$$\frac{1}{2\pi i}\int\limits_{L_1} \frac{æ_1 \ln(f_1(\zeta) + a)}{\zeta - z}\,\mathrm{d}\zeta = \frac{1}{2\pi i}\int\limits_{L_2} \frac{æ_2 \ln(f_2(\zeta) + a)}{\zeta - z}\,\mathrm{d}\zeta$$

$$\frac{1}{2\pi i}\int\limits_{L_1} \frac{æ_1 \ln(f_1(\zeta) - a)}{\zeta - z}\,\mathrm{d}\zeta = \frac{1}{2\pi i}\int\limits_{L_2} \frac{æ_2 \ln(f_2(\zeta) - a)}{\zeta - z}\,\mathrm{d}\zeta$$

hold, where $æ_1 \geq æ_2 > 0$, then $æ_1 = æ_2$.

From these theorems it follows that the stipulation of magnetic potentials, induced in a natural field and in a field cable, or potentials, induced at two situations of the cable, allows us uniquely to define the magnetic susceptibility. The uniqueness of determination of areas does not follow directly from the proved theorems, though, in conditions of the Theorem 1.3, this follows from the theorem of uniqueness for areas of identical constant density. Though Theorems 1.3 and 1.4 were proved for the inverse problem without taking account of demagnetization, for physical reasons, they are probably also valid even when demagnetization is accounted for. In this connection it is interesting to consider the problem of finding a unique solution of the inverse problem in practical situations. We shall illustrate this with examples. We shall consider an equivalent family of confocal ellipses in an earth field, directed along one of the axes of the ellipses. For one of them, with $æ = 0.3$, the direct MAM problem in Statement 1 was solved. The values calculated on a horizontal profile of $\mathrm{Im}(u_1(z))$ were approximated by one term of the sum (1.51) with accuracy 2% . Then for

$$u_1(z) = A_1 \left[1/(z - a_1) - 1/(z - b_1)\right]$$

the inverse problem for various values æ was solved.

From the theoretical results, it follows that at æ = 0.3, areas in equivalent families should coincide. In Figure 1.9 we see the domains for various values of æ from equivalent families. Areas with æ = 0.3 differ only within the accuracy of the method. The second example, the allocation of the unique solution of the inverse problem at the stipulation of potentials at two situations of a point source, is constructed similarly, and is illustrated in Figure 1.10. The real solution at æ = 0.26 is also determined rather well.

1.7. ON THE WEAK UNIQUENESS OF THE SOLUTION OF THE INVERSE PROBLEM

The concept of weak uniqueness for the TIP was introduced by Strakhov (1973), that is, for a class of areas with constant æ, the boundary of which is described by a given equation with undetermined coefficients. Weak uniqueness of the solution of the inverse problem takes place if, in a given external field, coefficients and magnetic susceptibility æ are uniquely determined. We shall show that in a 'slanting' field in the equivalent family $u_1(z)$ of form (1.22), an ellipse is the only area; that is, ellipses are a class of weak uniqueness of TIP. (This fact was established in another way by Blokh). Let the ellipse, filled with a substance with a constant magnetic susceptibility æ, and with semiaxes a and b ($a > b$) be located in a primary field H_0 (see Figure 1.2), where β is the angle between the major axis of an ellipse and the x-axis. The complex intensity of the ellipse $u_1(z)$ has form (1.22), and (as follows from (1.5))

$$\lambda = \frac{1 + 2\pi æ}{2\pi æ} \tag{1.65}$$

is a real number; the function $f(z)$ in this case has the form

$$f(z) = \left(\frac{z(a^2 + b^2)}{c^2} e^{-i\beta} \pm \frac{2ab}{c^2}\sqrt{z^2 e^{-2i\beta} - c^2}\right) e^{-i\beta}$$

and

$$f'(z) = e^{-2i\beta}\left(\frac{a^2 + b^2}{c^2} \pm \frac{2ab}{c^2}\frac{z e^{-i\beta}}{\sqrt{z^2 e^{-2i\beta} - c^2}}\right) e^{-i\beta} \tag{1.66}$$

From (1.66) we get

$$f'(\infty) = \frac{a-b}{a+b}e^{-2i\beta} = \alpha e^{-2i\beta}. \tag{1.67}$$

Substituting (1.66), (1.67) in (1.22), we obtain

$$u_1(z) = \frac{\lambda\overline{H} + \alpha e^{2i\beta}H}{\lambda^2 - \alpha^2 e^{4i\beta}} \frac{2ab}{c^2} e^{-2i\beta} \left(1 \pm \frac{2ab}{c^2} \frac{z}{\sqrt{z^2 - c^2 e^{2i\beta}}}\right). \tag{1.68}$$

As we see from (1.68), the focal distance c and angle β on $u_1(z)$ are uniquely determined. We shall prove that if the field (1.68) is generate by an ellipse with real parameters λ_1, a_1, b_1, the complex intensity of which is identically equal to $u_1(z)$, as represented by formula (1.68), then $\lambda = \lambda_1$, $a = a_1$, $b = b_1$. Hence, the relation

$$\frac{\lambda\overline{H_0} + \alpha e^{2i\beta}H}{\lambda^2 - \alpha^2 e^{4i\beta}} ab = \frac{\lambda_1\overline{H_0} + \alpha_1 e^{2i\beta}H}{\lambda_1^2 - \alpha_1^2 e^{4i\beta}} a_1 b_1 \tag{1.69}$$

holds; after simple transformations, we get the equation relative to λ_1:

$$\begin{aligned} &\lambda_1^2(\lambda\overline{H} + \alpha e^{2i\beta}H)ab - \lambda_1(\lambda^2 - \alpha^2 e^{4i\beta})\overline{H}a_1 b_1 - \\ &-\alpha_1^2 e^{4i\beta}(\lambda\overline{H} + \alpha e^{2i\beta}H)ab - \alpha_1 e^{2i\beta}H(\lambda^2 - \alpha^2 e^{4i\beta})a_1 b_1 = 0. \end{aligned} \tag{1.70}$$

The roots $\lambda_1^{(1,2)}$ of this equation are determined by the formulas

$$\lambda_1^{(1)} = \frac{a_1 b_1 \overline{H}(\lambda^2 - \alpha^2 e^{4i\beta})}{(\lambda\overline{H} + \alpha e^{2i\beta}H)ab} + \alpha_1 e^{2i\beta},\ \lambda_1^{(2)} = \alpha_1 e^{2i\beta}. \tag{1.71}$$

Obviously, $\lambda_1^{(2)}$ is a complex number, and $\lambda_1^{(1)}$ is real only at $a_1 = a$, $b_1 = b$ and is equal to λ. So, we have proved the following theorem.

Theorem 1.5. *There exists a unique ellipse, creating a field (1.68), and its parameters are determined by formula (1.68).*

1.8. ON THE UNIQUENESS OF THE SOLUTION OF THE INVERSE PROBLEM OF POTENTIAL THEORY IN A FIELD MAGNETIZATION METHOD USING A ROTATING FIELD (SQ-VARIATION)

In connection with the creation of a new kind of equipment — portable quantum magnetometers, characterized by high sensitivity and stability - it becomes possible to solve traditional problems of magnetometry using a new method, in a homogeneous complex of modern hardware together with methodical and interpretational development. We should emphasize that the development does not give only qualitative improvement of the method.

So simple introduction in a method of an artificial field magnetization using quantum magnetometers or variometers as a measuring element without a fundamental change in the technique of monitoring does not produce dramatic improvement of the method. The method of variations in its vectorial variant (Undezenkov, 1978) failed because of imperfections in the equipment (quartz variometers). The data of the variational method defined the size of the attitude(relation) of a residual magnetization (Kenigsberger factor) and coordinate of the centre of magnetic weights; and in MAM (variant of sounding) we obtained effective magnetic susceptibility and depth up to the top edge of the object. Thus, in each of these methods, the problem of definition or determination of the boundary of magnetic weights (nonlinear return problem of magnetic exploration) was not solved. With the application of the solution of the inverse problem (Martyshko, 1982) to the modular T-updating of the method of variations this has now become possible.

As field magnetization it is suggested to use a field of Sq-variations (Kamenskaya, 1966). The research of effects of a field magnetization with the help of these variations is of interest in connection with a large depth of penetration of the field (that is not so solid for an artificial source), regular occurrence of Sq-variations and their wide circulation. For depths in the first kilometers, the magnetization field of Sq-variations is practically similar this is not lack, as the advantage of MAM, connected with homogeneousity of magnetizating field, is lost with transition to study of depths further. The Sq-variations are generated by system of electric currents in the top layers of the atmosphere (ionosphere). General distribution of vectors of variations to the earth surface depends on the fact that the system of currents, adequate for Sq-variations, embraces an arc (30°) on a midday meridian and remains motionless in space between the Earth and Sun. The observer, rotating round this system, fixes during 24 hours all values of a vector of the intensity of the field, caused by the currents (Kamenskaya, 1966). Thus, we have a situation, similar to continuous moving of a source of a magnetizing field. Therefore, it is of interest to investigate questions of uniqueness of the solution of the inverse problem of magnetic exploration in the case of a field magnetization by a similar field, continuously changing in direction. The purpose is to show the possibility of determination of parameters of the object in the form of an elliptic cylinder from results of monitoring of effects of a field magnetization by a field of Sq-variations.

From Lavrentiev (1956 b) it follows that for the complete solution of the inverse problem, it is enough to carry out measurements of a secondary field at a continuously moving source of a field magnetizing field. In Tsirul'skii (1972), it is shown that if the magnetic potential induced at two lines of a source of a field magnetization of a field in the plane case is given (though one may be inhomogeneous), it allows us to determine uniquely the magnetic susceptibility and contour of the area, containing sources of the secondary field. Hence, in Tsirul'skii (1972), less rigid conditions, ensuring the complete solution of the inverse problem of a magnetic exploration (without account of demagnetization) are found. We shall show that in view of effect of demagnetization, the nonlinear inverse problem of magnetic exploration can be solved and for a similar field magnetizing field. We derived above the obvious kind of $u_1(z)$ for an ellipse in a similar arbitrarily oriented field ($u_3(z) = H$ – const):

$$u_1(z) = \frac{\lambda\overline{H} + \alpha e^{2i\beta} H}{\lambda^2 - \alpha^2 e^{4i\beta}} \frac{2ab}{c^2} e^{-2i\beta} \left(1 \pm \frac{z}{\sqrt{z^2 - c^2 e^{2i\beta}}}\right).$$

We set hereinafter that $\mathbf{H}$ is the field of variations and the function of time is $\mathbf{H} = \mathbf{H}(t)$. Suppose we have measured the values of $u_1(z)$ at some point z_0 in a rotating field $H(t)$ while the knot vector $\mathbf{H}$ describes a disk. As may be seen from formula (1.68), on $u_1(z)$ are uniquely determined the focal distance c and the angle β. We shall show that if the field is defined by an ellipse with the real parameters, α, λ, b, there is no other ellipse creating this field.

Proof. We shall assume the opposite, i.e. that there is an ellipse with real parameters, α_1, λ_1, b_1, whose complex intensity is identically equal to function $u_1(z_0, t)$, described by formula (1.68). Hence, the following equality should hold:

$$\frac{\lambda\overline{H} + \alpha e^{2i\beta} H}{\lambda^2 - \alpha^2 e^{4i\beta}} ab = \frac{\lambda_1\overline{H} + \alpha_1 e^{2i\beta} H}{\lambda_1^2 - \alpha_1^2 e^{4i\beta}} a_1 b_1$$

whence after simple transformations we get the equation involving λ_1

$$\lambda_1^2(\lambda\overline{H}+\alpha e^{2i\beta}H)ab-\lambda_1(\lambda^2-\alpha^2e^{4i\beta})\overline{H}a_1b_1-$$
$$-\alpha_1^2e^{4i\beta}(\lambda\overline{H}+\alpha e^{2i\beta}H)ab-\alpha_1e^{2i\beta}H(\lambda^2-\alpha^2e^{4i\beta})a_1b_1=0.$$

The roots of this equation $\lambda_1^{(1,2)}$ are set by the formulas

$$\lambda_1^{(1)}=\frac{a_1b_1\overline{H}(\lambda^2-\alpha^2e^{4i\beta})}{(\lambda\overline{H}+\alpha e^{2i\beta}H)ab}+\alpha_1e^{2i\beta},\ \lambda_1^{(2)}=\alpha_1e^{2i\beta}.$$

It is obvious that $\lambda_1^{(2)}$ is a complex number, $\lambda_1^{(1)}$ is real $(\forall H)$ only at $a_1=a$, $b_1=b$ and is equal to λ. Thus, is proved

Theorem 1.6. *There is a unique ellipse, creating a field (1.68), and its parameters are determined from formula (1.68).*

From the proved theorem follows the opportunity of finding the parameters of the ore deposit in the form of an elliptic cylinder using data of measurements of an abnormal magnetic field induced by a magnetizating rotating field. For the solution of the nonlinear inverse problem of magnetic exploration by the method of variations on monitoring at an abnormal point of secondary effects, the Sq-variations induced by a rotating field magnetizating field, according to the formula (1.68), it is necessary to know, at each moment, values of a field $u_1(z_0,t)$ and $u_3(z_0,t)$ (vector of geomagnetic variations $\mathbf{H}$).

To select the abnormal field $u_1(z_0,t)$ in the modular variant of a method of variations, we apply the differential circuit of monitoring: the values of the module of the complete vector T of the geomagnetic field in abnormal $T_a(t)$ and normal $(u_1=0)$ points $T_n(t)$ are simultaneously registered. The difference observed in result $T(t)=T_a(t)-T_n(t)$ contains the information on size of an abnormal field $u_1(z_0,t)$. Because of peculiarities of the modular variant of the method of variations in values of $\Delta T(t)$, there is an error caused by noncollinearity of the vectors T_a and T_n. This error may be estimated from the expression

$$\partial H_T=\frac{H_Z}{H_T}\partial H_Z+\frac{H_X}{H_T}\partial H_X+\frac{H_Y}{H_T}\partial H_Y. \tag{1.72}$$

Here ∂H_Y, ∂H_Z are the components of the vector of the geomagnetic field T and their variations accordingly.

Thus, from the analysis of expression (1.72), the component of a vector of a geomagnetic field and its variations $H(t)$ will necessarily be recorded at any abnormal point. That is, for reception of an abnormal field $u_1(z_0, t)$, it is necessary to observe variations of the module of a complete vector T in abnormal and normal points, and also to register variations of a component of a geomagnetic field in a normal point. Also, it is necessary to know direction cosines of the vector of the geomagnetic field at both kinds of points.

Let us examine an example of interpretation. On a practical example we shall illustrate the opportunity of determination of parameters of an anomaly forming body on data of measurements of a magnetic field, induced by a magnetizing field of Sq-variations (Martyshko and Pyankov, 1989). As an object of study, a quasi-two-dimensional anomaly of a submeridional extent MA-1 (Fig. 1.11) is chosen. The amplitude of the geomagnetic field in the epicentre of the anomaly is: $T_a = 63387$ nTl, $Z_a = 61090$ nTl, in a normal field $T_n = 60546$ nTl, $Z_n = 57590$ nTl. Observations of the vertical component of the geomagnetic field are carried out through a ferrozoned magnetometer MFD-4. In Fig. 1.11 the results of interpretation of an abnormal statistical field T on a secant latitudinal structure are illustrated. The interpretation involves a method of special points for revealing a two-dimensional interpretation of an observable field. As a result, we find that the western anomaly, on which variational measurements were conducted, is caused by a quasi-two-dimensional body of a bed with an angle of declination of about $30 - 40^o$.

Variational observation were carried out by the technique of synchronous measurements using highly stable proton magnetometers with an error of 0.05 nTl. One magnetometer was placed in the western epicentre of anomaly MA-1, another - at several kilometers from the first in a normal field. Measurements were made at intervals of 5 min. during a light day. The data are presented in Fig. 12. It may be seen that the error, caused by noncollinearity of the vectors T_a and T_n, is commensurable with a useful signal (~ 1 nTl) and essentially deforms the curve $u_1(z_0, t)$.

As the abnormal object is two-dimensional and has a submeridional extent, then, according to (1.72), it is possible to interpret a shaped curve as H_Z without entering large errors into results of selection. As the first approximation for interpretation of the data received at the analysis, the static field T was used. The interpretation was conducted by the trial and error method with use of data on the vector of geomagnetic variations $\mathbf{H} = \mathbf{H}(t)$ at each stage. As a result, an ellipse with the following parameters $a = 1250$ m, $b = 250$ m, æ $= 0.015$ knots SGSM was derived, angle $\beta = 30^o$ (almost vertical axis), depth up to the top edge $h = 750$ m.

Thus, from this interpretation, the determination of a unique solution of the nonlinear inverse problem of the method of field magnetization, the rotating field of Sq-variations on data of synchronous differential measurement of the module of a complete vector T of a geomagnetic field is shown. The offered approach — methodical basis for operative interpretation of magnetic anomalies has, in comparison with earlier existing methods, the following advantages:

a) With the help of a powerful source of a similar rotating magnetizating field (the Sq-variation) it is possible to increase the assessed depth of a method of field magnetization and to lower the labour involved in input;

b) Application of the new interpretational method allows us to determine the boundary and magnetic susceptibility of an abnormal object simultaneously. This was impossible previously.

We shall note also that the experimental data was submitted by Pyankov (Martyshko and Pyankov, 1989).

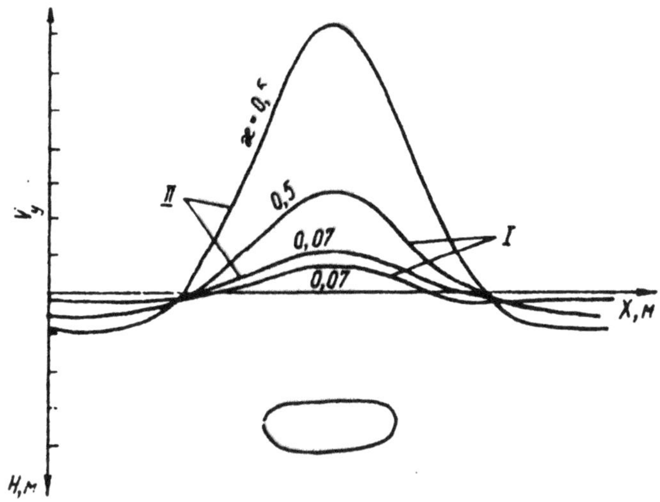

Figure 1.1

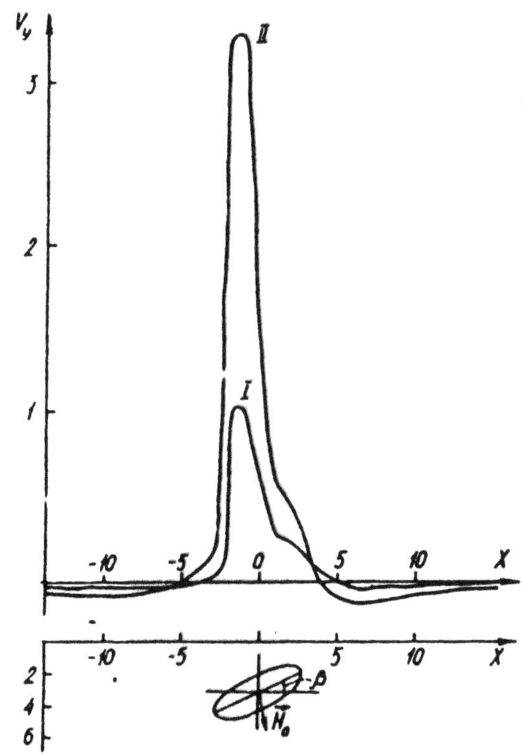

Figure 1.2

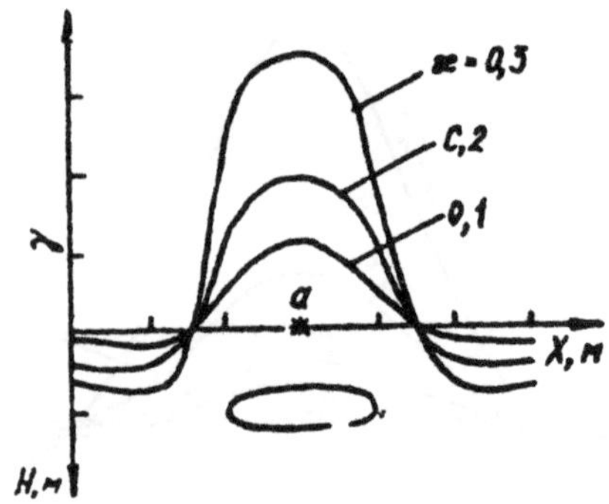

Figure 1.3

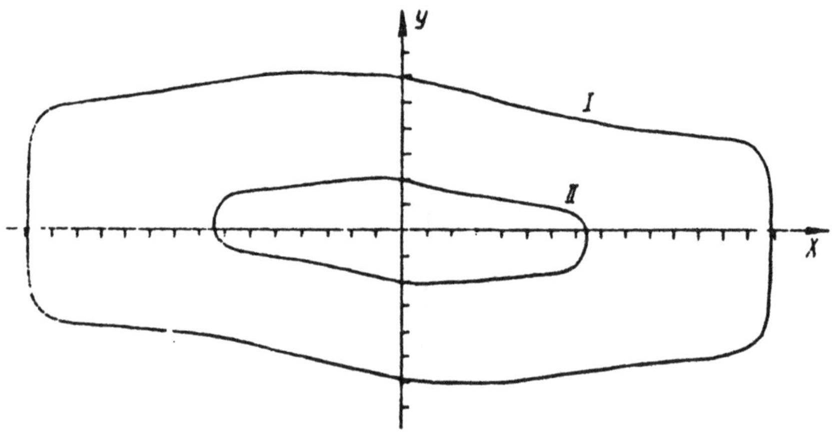

Figure 1.4

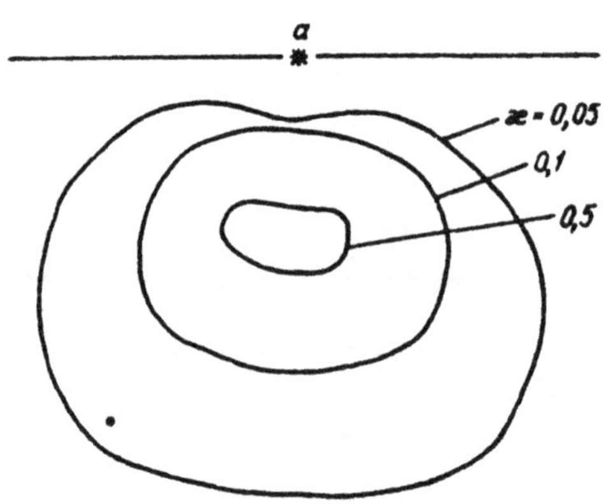

Figure 1.5

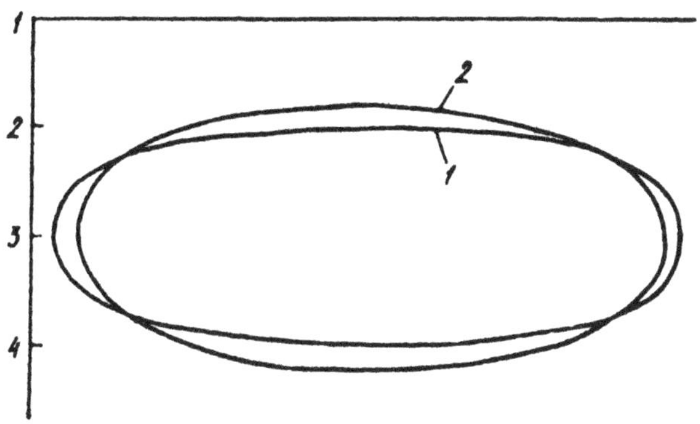

Figure 1.6

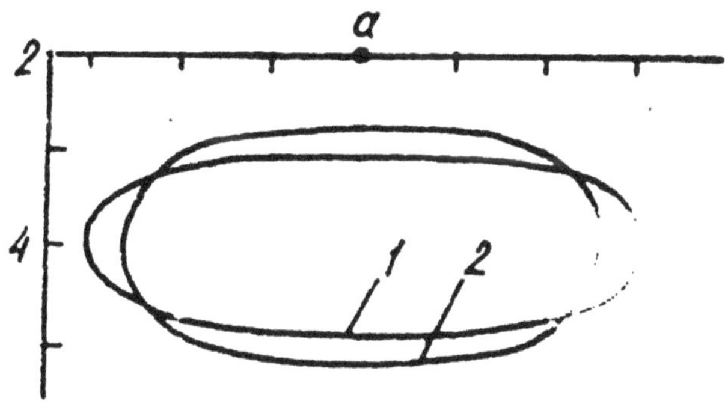

Figure 1.7

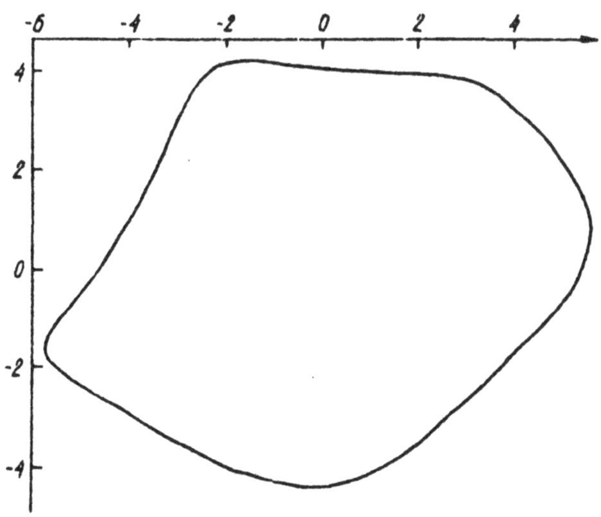

Figure 1.8

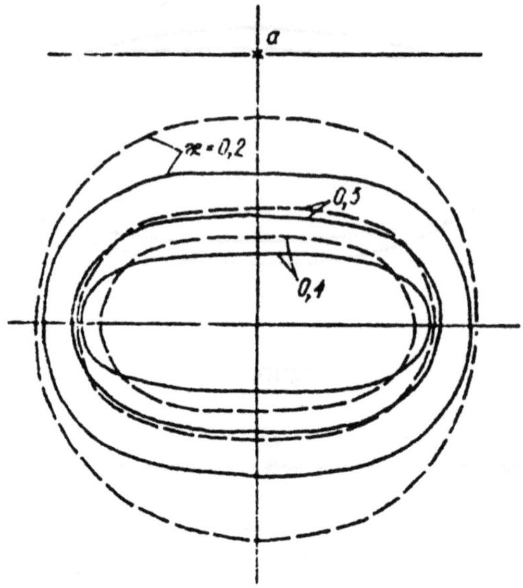

Figure 1.9

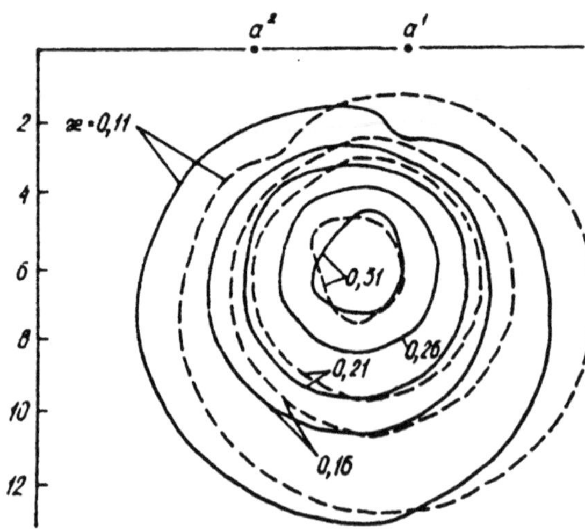

Figure 1.10

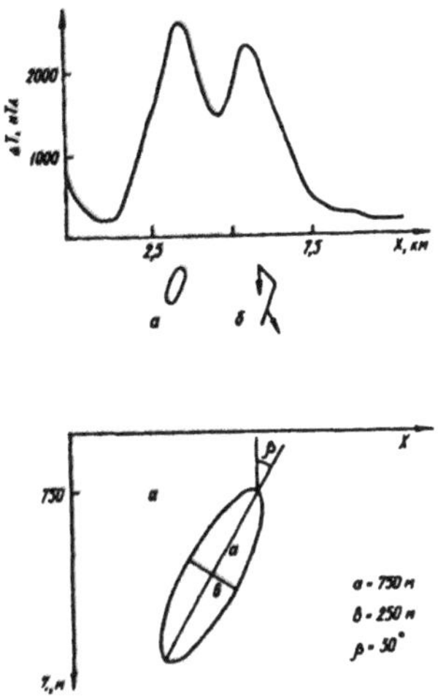

Figure 1.11

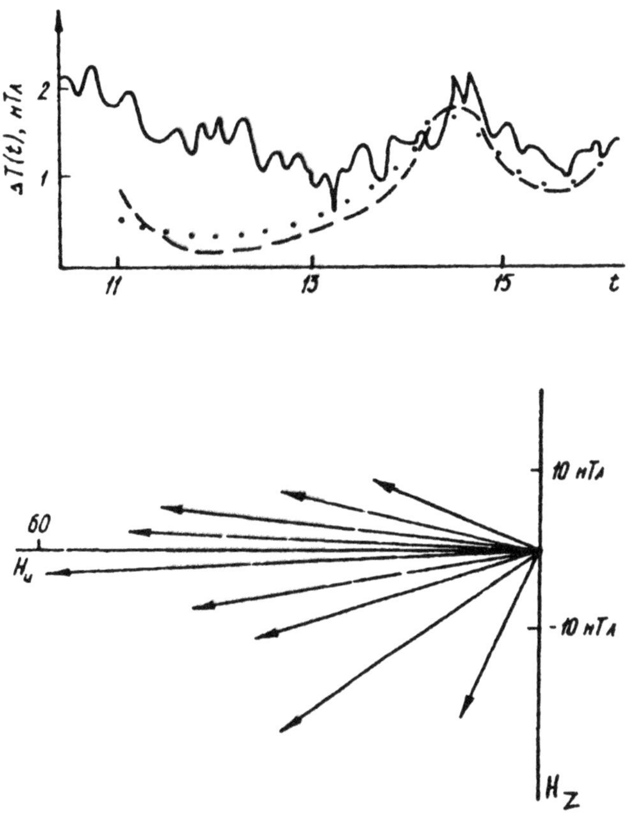

Figure 1.12

Chapter 2.

The effective algorithms of solution of direct and inverse three-dimensional problems of magnetic exploration taking account of demagnetization

In Chapter 1 there were investigated basic theoretical questions of two-dimensional problems: existence of equivalent families of solutions TIP, conditions, which were sufficient to determine the unique solution TIP, and also effective algorithms of the solution of inverse problems on the basis of the fast solution of direct problem in a parametrical class. From the practical point of view, the development of similar algorithms in the three-dimensional case is if an interest.

2.1. REDUCING THE DIRECT PROBLEM TO THE INTEGRAL EQUATION OF THE SECOND KIND

The classical method of solution of the direct three-dimensional problem of magnetic exploration in view of demagnetization consists in reducing the problem (0.1) to the integral equation of the second kind relative to magnetization of a body required density of a simple or double layer; then solving problem (0.1) (Vogel, 1963; Sharma, 1966; Sobolev and Belogolov, 1968; Voskoboynikov, 1973; Berdichevskii, 1974; and Blokh, 1975, 1987).

The first approach is more preferable to modeling fields from complex objects: few bodies with various values of magnetic susceptibility, cross-section with piecewise continuous change of æ. This approach was realized by Vogel (1963), Sharma (1966), Sobolev and Belogolov (1968) and other researchers. Blokh (1975, 1981, 1987), on the basis of the numerical solution of the integral equation for magnetization, had developed the effective apparatus of solution of direct problems of magnetic exploration and low-frequency inductive electric exploration, suitable for study of homogeneous and inhomogeneous, isotropic and anisotropic magnetic geological objects.

The second approach allows us to realize algorithms of solution of direct problems in parametrical classes. On the basis of it is possible to develop the methods of solution of inverse problems.

In the second case, when solution is searched as a double layer with the unknown density $\sigma(M)$, we have the following equation for $\sigma(M)$ (Voskoboynikov, 1973; Mayergoyz, 1979)

$$\sigma(M) - \frac{\nu}{2\pi} \int^{S} K(M,Q)\sigma(Q)\mathrm{d}S_Q = -\frac{\nu}{2\pi} W(Q) \tag{2.1}$$

where

$$S = \partial D, \quad M, Q \in S, \quad K(M,Q)\mathrm{d}S = \frac{(\rho, \mathbf{n})}{\rho^3}\mathrm{d}S = \mathrm{d}\omega. \tag{2.2}$$

Here ρ is the radius vector of the point Q relative to the point M; $\mathbf{n}$ is the unit vector of external normal to the surface S in the point Q; $\mathrm{d}\omega$ is the corporal angle, under which an element of a surface dS_q is seen from the point M

$$\nu = \frac{\mu_2 - \mu_1}{\mu_2 + \mu_1}. \tag{2.3}$$

Remark 2.1. In the assumption that S is a Lyapunov surface (Sobolev, 1962), it was proved, that equation (2.1) is the equation with a weak peculiarity; and we may apply Fredholm theory.

Remark 2.2. Closed double layer of constant density creates zero external and internal potentials and, hence, gives no contribution to intensity of field; therefore, it is possible to find $\sigma(M)$ within the limits of an additive constant (Voskoboynikov, 1973).

We shall note some properties of equation (2.1) (Filatov, 1969; Voskoboynikov, 1973). Least of the modulus eigenvalue of the integral operator of equation (2.1) is $\nu_1 = 1$. A unique eigenfunction φ_1 corresponds to the eigenvalue ν_1 indepently of the form of the surface S; φ_1 is a constant (this will be important for us further).

Numerical solution of equation (2.1) is rather hard because that $\nu = 1$ is an eigenvalue of equation (2.1). In this connection, the method of consecutive approximations for the equation for $\nu = 1$ fails; and for ν, close to 1, converges slowly. So, bad stability of the decision with respect to small errors of calculations takes place. Therefore, there is the problem of development of the algorithm of solution of the integral equation with an asymmetrical nucleus in a neighborhood of the first eigenvalue. For this purpose, it is necessary to allocate from the solution composed, containing set of the eigenfunctions appropriate to this number and to dispose it according to the requirements of a soluble problem. Voskoboynikov (1973) had offered to use instead of equation (2.1) the equation

$$\sigma(M) - \frac{\nu}{2\pi} \int^{S} \Big\{K(M,Q) - \frac{1}{|S|} \int^{S} K(M,Q)\,\mathrm{d}S_M\Big\}\sigma(Q)\,\mathrm{d}S_Q = \\ = -\frac{\nu}{2\pi}\Big\{W(M) - \frac{1}{|S|}\int^{S} W(Q)\,\mathrm{d}S_Q\Big\}. \tag{2.4}$$

For solution of problems of electric engineering, Mayergoyz (1979) had considered the equation

$$\sigma(M) - \frac{\nu}{2\pi} \int^{S} \sigma(Q)\left[K(M,Q) - \frac{2\pi}{|S|}\right] dS_Q = -\frac{\nu}{2\pi} W(M). \tag{2.5}$$

The solutions of equations (2.4) and (2.5) differ from the solution of (2.1) by a constant (hence, by virtue of Remark 2.2, they are equivalent, in the sense of the solution of problem (0.1) of equation (2.1)). The first is equal to the second eigenvalue of equation (2.1) $\nu_2 > 1$; therefore, their numerical solving has no mechanism of accumulation of an error, peculiar to equation (2.1). The proof of the facts formulated above about eigenvalues of equations (2.4) and (2.5) for the first equation was done by Voskoboynikov; for second we shall do it below.

2.2. SOLUTION OF THE OPERATOR EQUATION IN A NEIGHBORHOOD OF THE FIRST EIGENVALUE

We shall consider the operator equation

$$h - \nu K h = f \tag{2.6}$$

where h is known; f is a given element of a Gilbert space G; K is a completely continuous linear operator, which has at most one eigenvalue; ν is a parameter. Let in a general case $\{\nu_k\}_{k=1}^{\infty}$ be eigenvalues of the operator K, located by increasing of modules; φ_k be normalized eigenfunctions of the operator K appropriate to eigenvalues ν_k, i.e.

$$\varphi_k = \nu_k K \varphi_k, \quad \| \varphi_k \|^2 = (\varphi_k, \varphi_k) = 1. \tag{2.7}$$

Here φ_k^* are eigenfunctions of the conjugate operator K^*, hence,

$$(\varphi_k, \varphi_l^*) = 0, \text{ if } l \neq k. \tag{2.8}$$

Let the eigen subspace of the operator K, appropriate to eigenvalue ν_1, be as follows

$$G_0 = \{m\varphi_1\} \tag{2.9}$$

where m is any constant, and $G^{\perp}$ is the orthogonal complement to the complete space G, so that

$$G = G^{\perp} + G_0.$$

We shall introduce the operator

$$Lh = Kh - (1/\nu_1)\varphi_1(\varphi_1, h). \tag{2.10}$$

Relative to the operator L the following theorems hold.

Theorem 2.1. *Narrowings of the operator on a subspace G_0 is the operator of cancellation.*

Proof. From (2.9) and (2.10) we derive

$$Lm\varphi_1 = m\left[K\varphi_1 - \frac{1}{\nu_1}\varphi_1(\varphi_1, \varphi_1)\right].$$

whence, with the account of (2.7),

$$Lm\varphi_1 = 0. \tag{2.11}$$

Theorem 2.2. *Operators L cannot have eigenvalues, not equal to one of eigenvalues of the operator K. Eigenfunctions of the operator L are*

$$\Psi_k = \varphi_k - \alpha_k\varphi_k\nu_k/\nu_1$$

within the limits of a constant multiplier.

Proof. By the definition eigenfunctions of the operator L must satisfy the equation

$$\Psi - \nu L\Psi = 0 \tag{2.12}$$

for a certain value of the parameter ν. Substituting (2.10) in (2.12), we obtain

$$\Psi - \nu K\Psi = \frac{\nu}{\nu_1}\varphi_1(\varphi_1, \Psi). \tag{2.13}$$

This means that Ψ is a solution of inhomogeneous equation (2.13) also, the constant term of which belongs to G_0. By virtue of invariancy of G_0 (as of an own subspace) relative to the operator K, $\Psi \in G_0$; therefore, we shall search the solution in the form

$$\Psi = m\varphi_1 \tag{2.14}$$

where m is a factor to determine.

Let $\nu \neq \nu_k$. Then, for each constant term, equation (2.13) has a unique solution. Substituting (2.14) in (2.13), we obtain the equation for determination of m:

$$m(1 - \nu/\nu_1) = -m\nu/\nu_1.$$

Whence it follows that $m = 0$ and, hence, $\Psi = 0$.

Thus, the values $\nu \neq \nu_k$ cannot be eigenvalues of the operator L, and the theorem is proved.

We shall find eigenfunctions of the operator L. From above it follows that they satisfy the equations:

$$\Psi_k - \nu_k L \Psi_k = 0, \tag{2.12'}$$

$$\Psi_k - \nu_k K \Psi_k = -\frac{\nu_k}{\nu_1}(\varphi_1, \Psi_k) \tag{2.13'}$$

for some k.

Let $k \neq 1$. Then $(\varphi_1, \varphi_k^*) = 0$ (by virtue of (2.8)). Hence, the constant term of equation (2.12') satisfies conditions necessary and sufficient for existence of solution of this equation for values $\nu = \nu_k$. Repeating considerations for justification of (2.14), we find that $\Psi_k^0 = m\varphi_1$ is a particular solution of (2.13'). Adding to this solution the appropriate solution of homogeneous equation, we obtain the general solution

$$\Psi_k = m\varphi_1 + n\varphi_k \tag{2.15}$$

where m and n are constant factors, for determination of which we have the following relation from (2.14)

$$m\,[1 - \nu_k/\nu_1]\,\varphi_1 + n[\varphi_k - \nu_k K\varphi_k] = -\nu_k\varphi_1[m(\varphi_1, \varphi_1) + n(\varphi_1, \varphi_k)]/\nu_1.$$

This means that

$$m\,[1 - \nu_k/\nu_1] = -\nu_k[m + n\alpha_k]/\nu_1 \tag{2.16}$$

where $\alpha_k = (\varphi_1, \varphi_k)$.

We see that $m = -\alpha_k n \nu_k \varphi_1 / \nu_1$. Therefore, to within a constant multiplier, we have

$$\Psi_k = \varphi_k - \alpha_k \varphi_1 \nu_k / \nu_1. \tag{2.17}$$

Hence, the values $\nu = \nu_k$ for $k = 2, 3, \ldots$ are eigenvalues of the operator L, which correspond to eigenfunctions (2.17).

We shall consider now the case $k = 1$. The condition of existence of the solution of equation (2.12) in a general case is not researched.

The two variants are possible:

1) if $(\varphi_1, \varphi_1^*) \neq 0$, then for $\Psi_1 \neq 0$ equations (2.12') have no solution; hence, unique solution of equations (2.12) and (2.14') is $\Psi_1 = 0$;

2) if $(\varphi_1, \varphi_1^*) = 0$, the general solution (2.15) of equation (2.13') looks like (2.15); carrying out the proof, similar to the proof of Theorem 2.2, we see that $\Psi_1 = 0$. Hence, $\nu = \nu_1$ is not the eigenvalue of the operator L.

From the consideration above and Theorem 2.2 we have the following result.

Theorem 2.3. *The eigenvalue of the operator L least by the modulus is $\nu = \nu_2$.*

Let $G = L_2$ be the space of summable with the square Lebesgue functions, and K be the integral operator of equation (2.1). Then the operator L from (2.10) has the form:

$$Lh = \frac{1}{2\pi} \int^{S} h(Q) \left[K(M, Q) - \frac{2\pi}{|S|} \right] \mathrm{d}S. \tag{2.18}$$

In this case, $\varphi_1 = \sqrt{1/|S|}$.

From (2.18) it follows that the integral operator of equation (2.5) is a special case of operator (2.18); and for this operator the theorems proved above hold. Hence, the first eigenvalue of equation (2.5) is $\nu = \nu_2$, and it guarantees convergence at $\nu = 1$ of the method of consecutive approximations, which, according to (2.5), has the form

$$\sigma^n = \frac{\nu}{2\pi} \int_S \sigma^{n-1}(Q) \left[K(M, Q) - \frac{2\pi}{|S|} \right] dS_Q - \frac{\nu}{2\pi} W(M). \tag{2.19}$$

Or, in the operator form, the method of consecutive approximation is as follows

$$\sigma^n = \nu L \sigma^{n-1} + \sigma_0, \; \sigma_0 = -\frac{\nu}{2\pi} W(M)$$

that is

$$\sigma^n = \sum_{i=0}^{n} \nu^i L^i \sigma^0.$$

For the required density σ, we derive

$$\sigma = \lim_{n \to \infty} \sigma^n = \sum_{i=0}^{n} \nu^i L^i \sigma^0. \tag{2.20}$$

After determination of σ, it is possible to calculate the intensity of an induced field by the formula

$$\mathbf{H}(P) = \int_S \sigma(M) \, \mathrm{grad}_P \frac{(\rho, \mathbf{n})}{\rho^3} \, \mathrm{d}S \tag{2.21}$$

where $P \in \overline{D^-}$.

We shall note yet that the solution of the direct problem without the account of demagnetization can be obtained in the integral form (as a potential of a simple layer) (Sretenskii, 1946)

$$V(P) = -æ \int_S \frac{\partial W}{\partial n}(M) \frac{1}{\rho} \, \mathrm{d}S \tag{2.22}$$

whence

$$\mathbf{H}(P) = -æ \int_S \frac{\partial W}{\partial n} \, \mathrm{grad}_P \frac{1}{\rho} \, \mathrm{d}S. \tag{2.23}$$

Formulas (2.23) and (2.22), probably are more convenient than representations of V and $\mathbf{H}$ by volume integrals, as dimension of the space of integration is less. Besides, taking the appropriate parametrization of the surface S, these formulas are possible to consider as the nonlinear integral equations of the inverse problem in Statement 2.

The simple analysis shows that for calculation of $\sigma^n(M)$ by algorithm (2.19), on each step, we need smaller number of operations than in the similar algorithm for equation (2.4); therefore, we have prefered algorithm (2.19), which was realized by our program.

2.3. PARAMETRIZED ALGORITHMS OF THE SOLUTION OF DIRECT PROBLEMS

We shall consider a class of bodies, stellar relative to a certain internal point; this class is enough wide (from the practical point of view).

Let $\mathbf{r} = \mathbf{r}(\theta, \varphi)$ be the equation of a boundary of a body S in spherical coordinates.

The function vector $\mathbf{r} = \mathbf{r}(\theta, \varphi)$ is determined by the three scalar functions $x(\theta, \varphi)$, $y(\theta, \varphi)$, $z(\theta, \varphi)$:

$$\mathbf{r} = \{x, y, z\} \tag{2.24}$$

where

$$\begin{cases} x(\theta, \varphi) = r \sin\theta \cos\varphi, \\ y(\theta, \varphi) = r \sin\theta \sin\varphi, \\ z(\theta, \varphi) = r \cos\theta \end{cases} \tag{2.25}$$

and its derivatives $\mathbf{r}_\theta$, $\mathbf{r}_\varphi$ are determined by x_θ, y_θ, z_θ and x_φ, y_φ, z_φ respectively. They (as it follows from (2.25))have such representation

$$\begin{cases} x_\theta = r_\theta \sin\theta \cos\varphi + r \cos\theta \cos\varphi, \\ y_\theta = r_\theta \sin\theta \sin\varphi + r \cos\theta \sin\varphi, \\ z_\theta = r_\theta \cos\theta - r \sin\theta \end{cases} \tag{2.26}$$

$$\begin{cases} x_\varphi = r_\varphi \sin\theta \cos\varphi - r \sin\theta \sin\varphi, \\ y_\varphi = r_\varphi \sin\theta \sin\varphi + r \sin\theta \cos\varphi, \\ z_\varphi = r_\varphi \cos\theta. \end{cases} \tag{2.27}$$

From the vector product definition, we have

$$[\mathbf{r}_\theta \times \mathbf{r}_\varphi] = \left\{ \begin{vmatrix} y_\theta & z_\theta \\ y_\varphi & z_\varphi \end{vmatrix}, \begin{vmatrix} z_\theta & x_\theta \\ z_\varphi & x_\varphi \end{vmatrix}, \begin{vmatrix} x_\theta & y_\theta \\ x_\varphi & y_\varphi \end{vmatrix} \right\} \tag{2.28}$$

whence, in view of formulas (3.26), (3.27), after simple transformations, we derive:

$$[\mathbf{r}_\theta \times \mathbf{r}_\varphi] = \{r \cdot r_\varphi \sin\varphi + r^2 \sin^2\theta \cos\varphi - r \cdot r_\theta \cos\varphi \cdot \sin\theta \cdot \cos\theta,$$

$$-r \cdot r_\varphi \cos\varphi + r^2 \sin\varphi \cdot \sin^2\theta - r \cdot r_\theta \sin\theta \cdot \sin\varphi \cdot \cos\theta, \quad (2.29)$$

$$r^2 \sin\theta \cdot \cos\theta + r^2 \sin\theta \cdot \cos\theta + r \cdot r_\theta \sin^2\theta\}$$

$$|[\mathbf{r}_\theta \times \mathbf{r}_\varphi]| = \sqrt{\left[r^2(\theta,\varphi) + r_\theta^2(\theta,\varphi)\right]\sin^2\theta + r_\theta^2(\theta,\varphi) r(\theta,\varphi)}. \quad (2.30)$$

An orthonormal vector is as follows

$$\mathbf{n} = [\mathbf{r}_\theta \times \mathbf{r}_\varphi] \,/\, |[\mathbf{r}_\theta \times \mathbf{r}_\varphi]|, \;\; \rho = \left|r^0 - r\right|,$$

$$\mathrm{d}S = r(\theta,\varphi)\sqrt{\left[r^2(\theta,\varphi) + r_\theta^2(\theta,\varphi)\right]\sin^2\theta + r_\theta^2(\theta,\varphi)\mathrm{d}\theta\mathrm{d}\varphi}. \quad (2.31)$$

Substituting (2.24) - (2.31) into the integral in (2.5), we obtain:

$$\int^S \sigma(Q)\left[K(M,Q) - \frac{2\pi}{|S|}\right]\mathrm{d}S_Q = \int_0^\pi\int_0^{2\pi} \Phi(\theta,\varphi;\theta^0,\varphi^0)\mathrm{d}\theta\mathrm{d}\varphi = J(\theta^0,\varphi^0) \quad (2.32)$$

where

$$\Phi(\theta,\varphi;\theta^0,\varphi^0) = \left[\frac{r^2 \sin[\theta r - r^0(\cos\theta \cdot \cos\theta^0 +}{\sqrt{[(r^0)^2 + r^2}}\right.$$

$$\frac{+\sin\theta \cdot \sin\theta^0 \cos(\varphi - \varphi^0))] - r r_\varphi r^0 \sin\theta^0 \cos(\varphi - \varphi^0)}{-2rr^0(\cos\varphi^0 \cdot \sin\theta^0 \cdot \cos\varphi \cdot \sin\theta +} \quad (2.33)$$

$$\frac{[\sin\theta \cdot \cos\theta^0 - \cos\theta \cdot \sin\theta^0 \cdot \cos(\varphi - \varphi^0)]}{+\sin\varphi^0 \cdot \sin\theta^0 \cdot \sin\varphi \cdot \sin\theta + \cos\theta^0 \cdot \cos\theta]^3}$$

$$-\frac{2\pi}{|S|}\left\|[\mathbf{r}_\theta \times \mathbf{r}_\varphi]\right\|\Big]\,\widetilde{\sigma}.$$

The integral $J(\theta^0, \varphi^0)$ is calculated by the formulas of the best degree of accuracy, which for 2π periodic functions $g(x)$ look as follows

$$\int_0^{2\pi} g(x)\mathrm{d}x \approx \frac{2\pi}{N}\sum_{k=1}^{N} g\left(\alpha + (k-1)\frac{2\pi}{N}\right). \tag{2.34}$$

Here $0 \le \alpha \le \frac{2\pi}{N}$. So, we have

$$J(\theta^0, \varphi^0) \approx \left(\frac{2\pi}{N}\right)^2 \sum_{k=1}^{N_1}\sum_{i=1}^{N} \Phi(\theta_k, \varphi_i) \tag{2.35}$$

where θ_k, φ_i are the knots with the equidistant step $h = 2\pi/N$, and $\theta_1 = \varphi_1 = \alpha$.

Thus, the problem of calculation of superficial integral in (2.5) is reduced to calculation of a double integral, for the numerical account of which the knots are independent of a kind of function $r(\theta, \varphi)$ (i.e., from the form of a surface). By the method described above it is possible to solve the direct problem for bodies with piece-smooth boundary also; thus, $r(\theta, \varphi)$ is set by the table. We shall consider, taking an example of a parallelepiped, how to calculate $\mathbf{n}$ and ds. Let $Ax + By + Cz + E = 0$ be the equation of one of sides in Cartesian coordinates; then it is possible to determine the normal vector to this side

$$\mathbf{N} = \{A, B, C\}. \tag{2.36}$$

From here we have

$$\mathbf{n} = \mathbf{N}/\left|\mathbf{N}\right|, \ \mathrm{d}s = \left|\mathbf{N}\right|\mathrm{d}\theta\mathrm{d}\varphi$$

where

$$|\mathbf{N}| = \sqrt{A^2 + B^2 + C^2}.$$

We shall note that in the case when a point M and an element of the surface ds belong to one plane, dω from formula (2.2) is equal to zero.

Therefore, for polyhedrons in formula (2.35), the summation will be carried out only by points, not belonging to the side, where the point M lays.

2.4. EXAMPLES OF MATHEMATICAL MODELING

We shall consider the operation of the algorithm using examples.

Example 1. We shall notice that for an ellipsoid of rotation in a field of a point source potential as a number(line) on spherical functions is known. For the purpose of estimation of accuracy of the algorithm, we consider the ellipsoid with the semiaxes $a = b = 5$, $c = 1$. The values of potential, obtained by the described method were done with the relative accuracy 0.001. The maximum deviation was 1.5% (Figure 2.1).

Example 2. Under the contract with Kustanay geological exploration expedition, mathematical modeling of data of square sounding MAM for a parallelepiped was carried out. Curves of sounding and card of isolines H_Z in coordinate planes $(DR/2, X)$, $(DR/2, Y)$, where $DR/2$ is a half of part of field magnetization square framework, H_X and H_Y are situated in the plane (x, y), were constructed (see, Martyshko *et al.*, 1988). It was shown that for definition of the depth up to the centre (and for a horizontal position of coordinates of the centre) it is enough to measure H_Z. However, for determination of sizes of the body and the direction of fall, it is necessary to measure H_X, H_Y. Thus, the cards of isolines H_X, H_Y in the plane (x, y) give a good representation of the horizontal sizes of a body; the poles of isolines correspond to the ends of the body, and the point of bridging corresponds to the centre at a horizontal position or to the centre of the top edge for the inclined position.

In Figure 2.2 the results of modeling for the horizontal parallelepiped with the parameters $æ = 0.1$, the coordinate of centre $x = y = 0$, $z = 5$, the ratio of lengths of edges $10 : 5 : 1$ are represented.

We shall note also that the results of mathematical modeling are well coordinated with the data of physical modeling (Martyshko *et al.*, 1985).

2.5. SOLUTION OF THE INVERSE PROBLEM IN VIEW OF DEMAGNETIZATION BY THE METHOD OF THE QUASI-EQUIVALENT ELLIPSOID

The essential progress in the field of the solution of two-dimensional problems of gravic exploration and magnetic exploration without the account of demagnetization effect is managed to be reached using the apparatus of the theory of functions of complex variable, with the help of allocation of an extensive class of potentials, for which the inverse problem is reduced to a finite system of nonlinear equations relative to the parameters of the required area. The advantages of such approach were remarked by Tsirul'skii and Nikonova (1975), and by Nikhonova and Tsirul'skii (1978). Unfortunately, in the three-dimensional case it is not possible to obtain similar classes of potentials. We shall remind that already in two-dimensional case another algorithm of the solution of the inverse problem 1 with use of nonlinear programming on the basis of the fast parametrized solution of the direct problem was developed. In this chapter, we develop a similar algorithm in the three-dimensional case.

For the solution of the inverse problem 1 in the class of stellar bodies it is convenient to use the following representation for the intensity H, which can be obtained from (2.21):

$$\mathbf{H} = \int_0^{\pi}\int_0^{2\pi} \widehat{\sigma}(r(\theta,\varphi),\theta,\varphi)\psi(r(\theta,\varphi),\theta,\varphi)\mathrm{d}\theta\mathrm{d}\varphi. \tag{2.37}$$

Here, $\mathbf{H}$ is any of the intensities H_X, H_Y, H_Z.

The kind of function Ψ is determined by relations (2.36), and $\widehat{\sigma}$ (as was shown above) is determined as a result of the numerical solution of the integral Fredholm equation of the second kind. In this connection, relation (2.37) may be considered as the operator equation of the first kind

$$A\mathbf{r} = \mathbf{H} \tag{2.38}$$

relative to the function $r(\theta,\varphi)$ with the implicitly given operator A.

In this case, the solution of the inverse problem may be reduced to minimization of the functional

$$f(\gamma) = \sum_{i=1}^{n} \mid H(M_i) - \widetilde{H}(M_i; r(\gamma) \mid^2 \tag{2.39}$$

by the parameters of the required function $r(\theta,\varphi)$, where $\mathbf{H}$ is the right-hand side (2.37); $\{M_i\}$ are the knots of some carrier S_0: $S_0 \cap \overline{D} = \emptyset$, where the values of magnetic field $H(M_i)$ are known (by results of supervision or selection); $r(\gamma) = r(\theta, \varphi; \gamma)$, $\gamma = (\gamma_1, \gamma_2, \dots, \gamma_N)$ is the vector of parameters which determines r.

As was mentioned by (Strakhov, 1974; Tsirul'skii and Nikonova, 1975; and by Nikhonova and Tsirul'skii, 1978), for development of algorithms of the solution of inverse problems by methods of nonlinear programming it is extremely important to choose an effective parametrization of a class, where the problem is solving. The problem of $r(\theta, \varphi)$ choice is especially urgent for the solution of the inverse MAM problem 1, as for the minimization of functional (2.39) it is necessary to solve the integral equation several times; this is connected with large expenditure of machine time. The probable parametrizations of $r(\theta, \varphi)$ where taken as a piece of a double Fourier series or as spherical functions. In this case, the number of parameters, which is possible to determine for minimization at acceptable time of account $N \sim 10$; however, 10 terms of a the series is not enough for the description of complex objects. On the other hand, the ellipsoid simulates ore bodies rather satisfactorily, and their boundary may be described by 9 parameters.

The function $r(\theta, \varphi)$ for an ellipsoid has the form

$$r(\theta, \varphi) = \frac{abc}{\sqrt{(a^2 \sin^2 \varphi + b^2 \cos^2 \varphi) c^2 \sin^2 \theta + a^2 b^2 \cos^2 \theta}} \tag{2.40}$$

where a, b, c are the semiaxis of the ellipsoid, being subject to determination as the solution of the inverse problem. In the general case, we have add six more parameters: x_0, y_0, z_0 the shift of the centre of the ellipsoid relative to the origin and α, β, γ the Euler angels which determine axes of the ellipsoid.

Thus, if we need to search for the solution of the inverse problem in form (2.40), the number of required parameters is 9, and these parameters may be defined by minimization of functional (2.39) on the computer. The algorithm stated above of the solution of the inverse problem was realized on the computer; the functional (2.39) was minimized by the modified generalized Powell method (Himmelblau, 1975). The algorithm allows us to find (given the value of æ) an ellipsoid by one of intensities of an abnormal magnetic field, which gives the minimum of functional (2.39) on the set of all ellipsoids with the magnetic susceptibility æ, i.e., find the quasi-solution of equation (2.37).

Thus, with the help of the algorithm described above, changing the value of parameter æ, it is possible to construct the family of quasi-equivalent ellipsoids. We shall notice that we solve the inverse problem twice approximately: 1) we search the solution as an ellipsoid (geometrical approximation);

2) by the method of minimization, we find parameters of an ellipsoid approximately. Therefore, the distinction between the real solution of the inverse problem and the approximate one can be, generally speaking, rather great. It is clear, that it is the more, the more differs the real object from the ellipsoid.

2.6. THEORETICAL AND PRACTICAL EXAMPLES OF REALIZATION OF THE ALGORITHM OF THE 'EQUIVALENT ELLIPSOID'

Example 3. The problem "of restoration of ellipsoid": for the ellipsoid with æ $= 0.1$ and the semiaxes $a = b = 5$, $c = 1$, the centre of which is on the depth 6, values of **H** in the knots of a flat grid on a day time surface were calculated; then the calculated values **H** were used for solution of the inverse problem by the method stated above. The result of solution (ellipsoid with the semiaxes $a = b = 4.92$, $c = 0.95$), received for 10 iterations, is quite satisfactory (Figure 2.3 1 – exact solution, 2 – approached solution of the inverse MAM problem, 3 – solution of the inverse problem without the account of demagnetization). The maximum divergence between **H** the approached and exact is 1.5% .

Example 4. We shall illustrate now the opportunity of application of developed algorithm for determination of the unique solution of the inverse problem for complexitation of the given magnetic exploration and MAM. We shall consider the family of confocal ellipsoids (equivalent in the earth field). For one of them (æ $= 0.1$), the direct problem 1 was solved, and then by the method stated above for various æ the family of quasi-equivalent ellipsoids was constructed. The results are brought on Figure 2.4: the continuous line designates sections by coordinate planes of confocal ellipsoids; the pointed line designates the families of quasi-solutions. The ellipsoids with æ $= 0.1$ coincide minimally. Their sections are designated by number 3 in the figure (we shall note, that the cross-sections done by the plane $z = 0$ of an ellipsoid of the quasi-solution of the inverse MAM problem at æ $= 0.1$ are practically coincided). The letter R marks the position of the centre of the framework; the points are their projections in the plane $z = 0$.

Practical examples.

1. The experiment-methodical work on a testing of dipole variant of the method of an artificial field magnetization were spent by Chelyabinsk geo-

logical exploration expedition on Glubochenskaya anomaly. On Figure 2.5 the plan of isodynamic lines of this anomaly is seen, and the position of the field magnetization framework is marked.

The measurements of horizontal component of the magnetic field H_X, H_Y were carried out only on structures 1 and 2; that is insufficiently for use of the given data for solution of the three-dimensional inverse MAM problem.

Taking into account the geological data, Glubochenskaya anomaly is formed by an ore body (on Figure 2.6 we see the section by structure 2) with a magnetic susceptibility æ = 0.1. For solution of the inverse problem, we have used the magnetic exploration of the area. The method described above constructs the family of ellipsoids 'quasi-equivalent' in an earth field with various values of æ; for each of them values of H in the field of the framework (the direct MAM problem was solved), the observed curve will be best approximated for the ellipsoid with æ = 0.1 (Figure 2.6), which will satisfactorily approximate the ore body.

2. In Figure 2.7 we see the plan of isodynamic lines of the Bogdanovskii anomal zone N 3. The field magnetization framework with the sizes 400×400 m was established on the flank of the anomaly; and on structure 1 the values of H_X and H_Y were measured. The family of quasi-equivalent (in the earth field) ellipsoids was constructed similarly, and for each of them the direct MAM problem was solved. As we see from Figure 2.8 the observed values will be best approximated by the ellipsoid with æ = 0.15 (on Figure 2.8 it is the pointed curve); determined data the ore body by the drilling (on Figure 2.8 is shaded) with æ = 0.15 is well approximated by this ellipsoid.

The examples above mentioned illustrate an opportunity of use of the developed algorithms and programs for interpretation of practical anomalies and opportunity of effective determination of the unique solution (i.e. magnetic susceptibility and geometrical parameters of a body) at a complexitation of MAM data and magnetic exploration or framework given MAM at two situations (better to complex data in a field of two frameworks, in order to avoid the influence of residual magnetization). The method of quasi-equivalent ellipsoid was considered only as one of the possible approaches of solution of the inverse problem on the computers with speed about 10^6 operations per second.

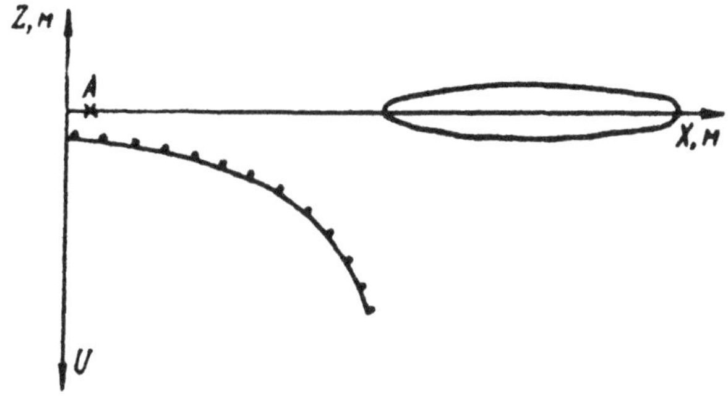

Figure 2.1

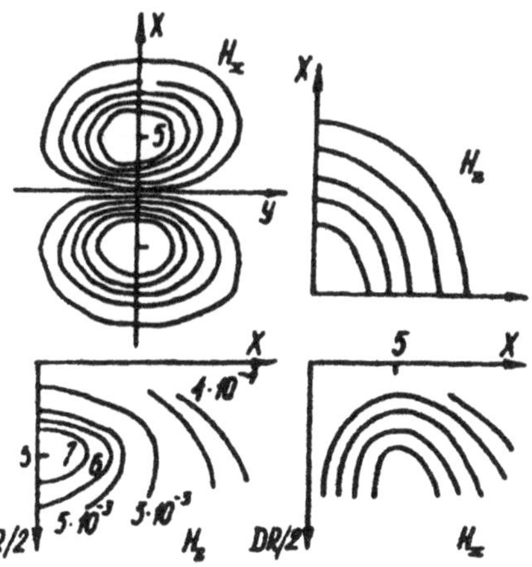

Figure 2.2

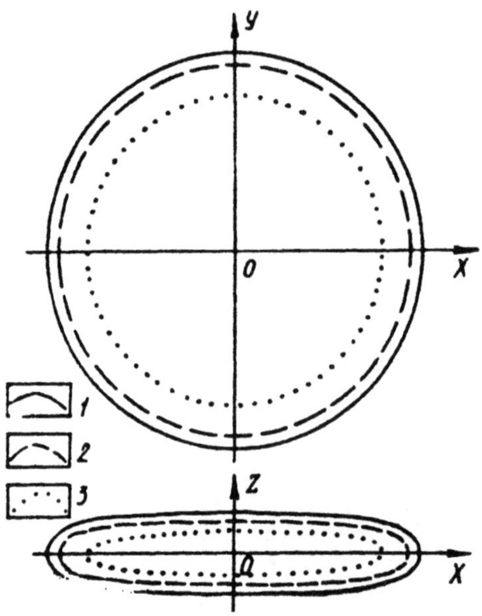

Figure 2.3

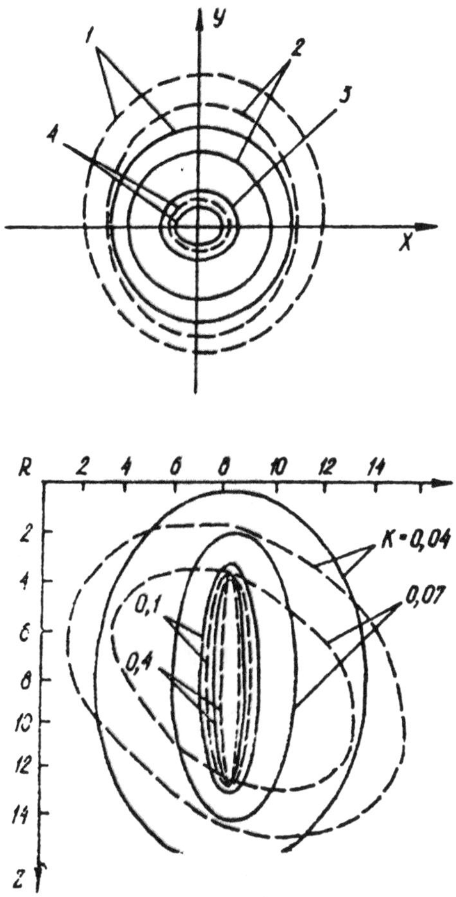

Figure 2.4

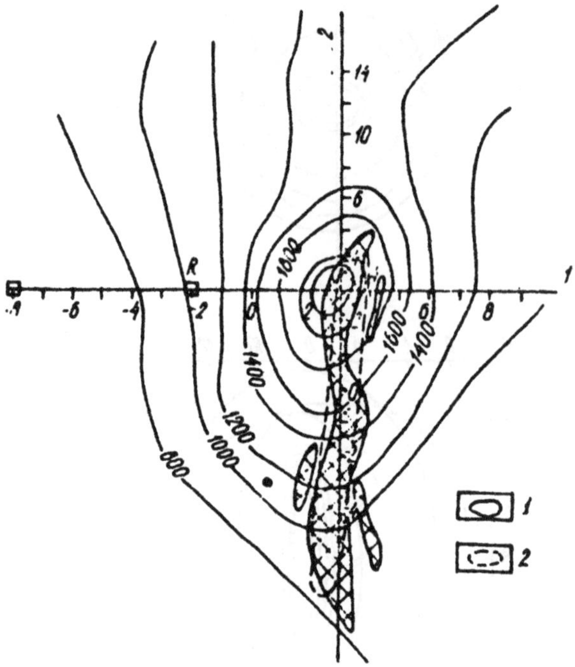

Figure 2.5

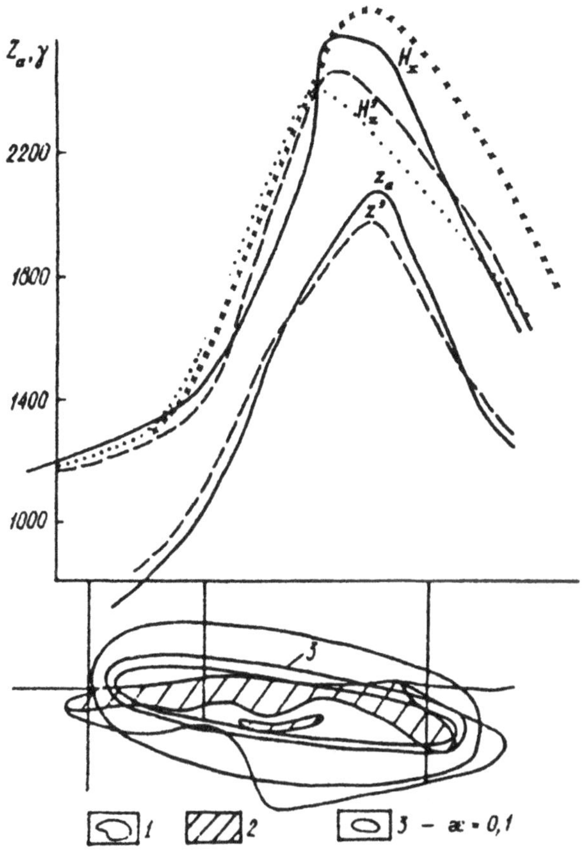

Figure 2.6

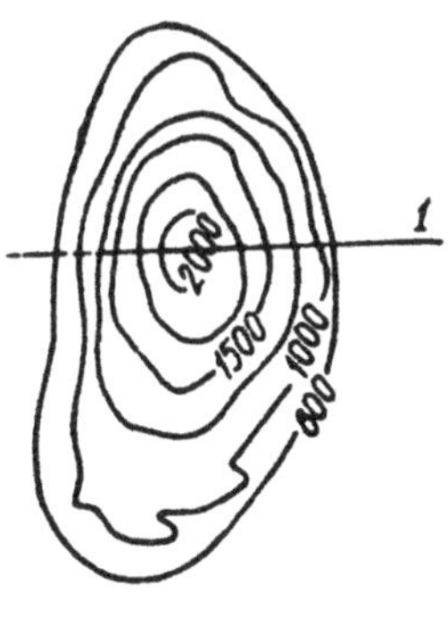

Figure 2.7

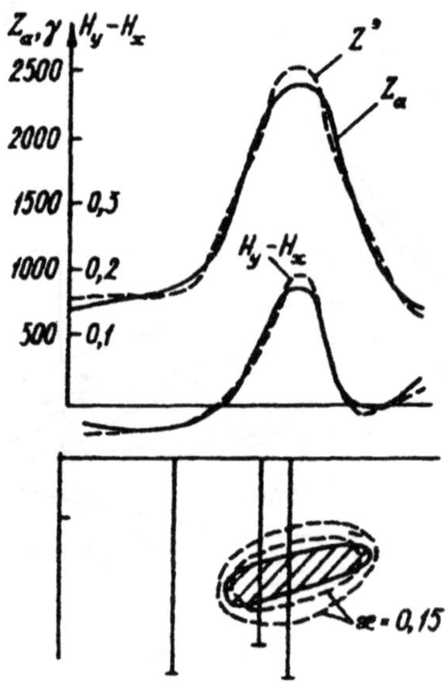

Figure 2.8

Chapter 3.

Mathematical theory and algorithms of the solution of three-dimensional inverse problems of electric exploration with a direct current

Interpreting the data of electric exploration often involves considerable difficulties: in the general case, the direct problem cannot be solved explicitly and the inverse problem is therefore reduced to an operator equation of the first kind with an operator defined not explicitly. Above, an algorithm had been suggested for the solution of a similar equation for problems of artificial magnetization which are equivalent mathematically to problems of electric exploration. A method of nonlinear programming based on a parametrized solution of the direct problem was used for this purpose. This algorithm, however, involves substantial use of computer time, because at each minimization step one has to solve repeatedly an integral equation. The algorithm has therefore been realized only for the class of ellipsoids. It is of interest to find equations with an explicitly defined operator for an solution of the inverse problem of electric exploration.

For the theoretic inverse problem (TIP) of electric exploration, where from an anomalous electric potential given explicitly one has to find an equivalent family of bodies with various conductivities which generate this potential in the field of a specific source, such an equation can be derived. Note that the solution of TIP is of interest not only as a realization of one stage in two-stage interpretation method (Nikonova and Tsirul'skii, 1978), but also because it allows us to constructing geologically meaningful equivalents of various classes of singular sources.

In this chapter we derive simpler equations (with explicitly defined operators) of the theoretical inverse problem of electric exploration and constructs algorithm of their solution for arbitrary potential classes.

3.1. PROBLEM STATEMENT

Let D be a one-connected domain of the Euclidean space R^3 of conductivity σ_2, which is located in a medium with conductivity σ_1 ($\sigma_1, \sigma_2 = \text{const}$); W be the electric potential of external sources in the absence of an inclusion (exciting potential); and V_1 and V_2 be the external and internal potentials of the conducting inclusion D. The potentials V_1, V_2, and W are harmonic functions, i.e.,

$$\Delta V_1 = 0 \; in \; D^-, \quad \Delta V_2 = 0 \quad \text{and} \quad \Delta W = 0 \in D. \tag{3.1}$$

In addition, at the boundary D (the surface S), the following relations take place:

$$V_1 = V_2 \tag{3.2}$$

$$\sigma_2 \frac{\partial V_2}{\partial n} - \sigma_1 \frac{\partial V_1}{\partial n} = (\sigma_1 - \sigma_2) \frac{\partial W}{\partial n}. \tag{3.3}$$

The solution of the direct problem of electric exploration is reduced to finding the potentials V_1 and V_2 from conditions (3.1)–(3.3). The inverse problem may be stated as follows: to find the desired domain from a given function V_1 if the boundary conditions (3.1)–(3.3) hold.

Remark 1. Similar problems (with substitution of magnetic potentials by electric ones and magnetic permeability μ for σ) arise when interpreting the data of the method of artificial magnetization (MAM).

3.2. INTEGRAL EQUATIONS OF THE THEORETICAL INVERSE PROBLEM OF ELECTRIC EXPLORATION AND THE METHOD OF ARTIFICIAL MAGNETIZATION

We shall call an inverse problem theoretical if V_1 is given explicitly (a situation arising, for example, after fitting the observed data to the potentials of singular sources).

The following expressions hold for the functions V_1 and V_2 (Sobolev and Belogolov, 1968):

$$\frac{1}{4\pi}\int\limits_S \left[V_1(P)\frac{\partial\frac{1}{r}}{\partial n} - \frac{\partial V_1}{\partial n}\frac{1}{r}\right] \mathrm{d}S = \begin{cases} \mathbf{V}_1(P_0),\, P_0 \in R^3/\overline{D}, \\ 0,\, P_0 \in D \end{cases} \tag{3.4}$$

$$\frac{1}{4\pi}\int\limits_S \left[\frac{\partial V_2}{\partial n}\frac{1}{r} - V_2(P)\frac{\partial\frac{1}{r}}{\partial n}\right] \mathrm{d}S = \begin{cases} \mathbf{V}_2(P_0),\, P_0 \in D, \\ 0,\, P_0 \in CD \end{cases} \tag{3.5}$$

where $\mathbf{n}$ is the external normal to S, $P(x,y,z) \in S$; $P_0(x_0,y_0,z_0) = P_0$, and $r = \sqrt{(x-x_0)^2+(y-y_0)^2+(z-z_0)^2}$.

From (3.4) and (3.5), in view of (3.2), it follows that

$$\frac{1}{4\pi}\int\limits_S \left[\frac{\partial V_2(P)}{\partial n} - \frac{\partial V_1(P)}{\partial n}\right]\frac{\mathrm{d}S}{R} = \begin{cases} \mathbf{V}_1(P_0),\, P_0 \in R^3/\overline{D}, \\ V_2(P_0),\, P_0 \in D. \end{cases} \tag{3.6}$$

From (3.3) it is easy to obtain

$$\frac{\partial V_2}{\partial n} - \frac{\partial V_1}{\partial n} = \frac{\sigma_1 - \sigma_2}{\sigma_2}\left(\frac{\partial W}{\partial n} + \frac{\partial V_1}{\partial n}\right) \tag{3.7}$$

and

$$\frac{\partial V_2}{\partial n} - \frac{\partial V_1}{\partial n} = \frac{\sigma_1 - \sigma_2}{\sigma_1}\left(\frac{\partial W}{\partial n} + \frac{\partial V_2}{\partial n}\right). \tag{3.8}$$

Substituting (3.8) into (3.6), we have

$$\begin{aligned} \mathbf{V}_1(P_0) &= \frac{1}{4\pi}\frac{\sigma_1-\sigma_2}{\sigma_1}\int\limits_S \left(\frac{\partial W(P)}{\partial n} + \frac{\partial V_2(P)}{\partial n}\right)\frac{\mathrm{d}S}{r} = \\ &= \frac{1}{4\pi}\frac{\sigma_1-\sigma_2}{\sigma_1}\int\limits_S (W(P)+V_2(P))\frac{\partial\frac{1}{r}}{\partial n}\,\mathrm{d}S. \end{aligned} \tag{3.9}$$

and, hence, by virtue of (3.2), we have (Martyshko, 1986):

$$\mathbf{V}_1(P_0) = \frac{1}{4\pi}\frac{\sigma_1 - \sigma_2}{\sigma_1} \int\limits_S (W + V_1)\frac{\partial \frac{1}{r}}{\partial n}\,\mathrm{d}S. \tag{3.10}$$

Relation (3.10) may be considered as an equation for defining the boundary S. Substituting (3.7) into (3.6), we obtain

$$\mathbf{V}_1(P_0) = \frac{1}{4\pi}\frac{\mu_1 - \mu_2}{\mu_1} \int\limits_S \left(\frac{\partial W(P)}{\partial n} + \frac{\partial V_1(P)}{\partial n}\right)\frac{\mathrm{d}S}{r} \tag{3.11}$$

and, hence,

$$\mathbf{H} = \operatorname{grad} \mathbf{V}_1 = \frac{1}{4\pi}\frac{\mu_1 - \mu_2}{\mu_1} \int\limits_S \left(\frac{\partial W}{\partial n} + \frac{\partial V_1}{\partial n}\right) \operatorname{grad}\frac{1}{r}\,\mathrm{d}S \tag{3.12}$$

which are the integral equations of the theoretical inverse problem of MAM.

The case of $\sigma_2 = \infty$ (ideal conductor).

This is of an interest as the extreme case, because at $\sigma_2 = \infty$ we have the minimal domain of the equivalent family of bodies creating for various conductivities σ_2 the same potential $\mathbf{V}_1$. Here, from (3.3), we obtain

$$\left.\frac{\partial V}{\partial n}\right|_S = -\left.\frac{\partial W}{\partial n}\right|_S. \tag{3.13}$$

Then, from (3.1) and (3.13), it follows that

$$V_2 = -W + \mathrm{const}$$

i.e., by virtue of (3.2), we have

$$(V_1 + W)\,|_S = \mathrm{const}. \tag{3.14}$$

Relation (3.14) may be considered as the equation defining the surface of an ideal conductor (which is known to be the surface of level for the total potential $U = V_1 + W$), i.e., as the equation of TIP.

It is important to note that TIP equations will have the same form if we formulate the problem for a set of bodies. Integration will be done along all boundaries.

3.3. INTEGRAL TRANSFORMATIONS OF ELECTROMAGNETIC FIELDS

Relation (3.10) is an equation for the boundary S. However, both relations may be viewed as an improvement of formulas (3.4) and (3.5) for the electric potentials of the domain D. Of an interest is the question of whether representation (3.10) for V_1 is a characteristic function.

Assume that a harmonic function V_3 exists everywhere outside D, so that

$$\mathrm{V}_3(P_0) = \frac{1}{4\pi}\frac{\sigma_1 - \sigma_2}{\sigma_1}\int_S [W + V_3]\,\frac{\partial \frac{1}{r}}{\partial n}\,\mathrm{d}S. \tag{3.10}$$

Then we have

$$V_1(P_0) - \mathrm{V}_3(P_0) = \frac{1}{4\pi}\frac{\sigma_1 - \sigma_2}{\sigma_1}\int_S (V_1 - V_3)\frac{\partial \frac{1}{r}}{\partial n}\,\mathrm{d}S$$

everywhere outside D, i.e.,

$$F(P_0) = \frac{1}{4\pi}\frac{\sigma_1 - \sigma_2}{\sigma_1}\int_S F\frac{\partial \frac{1}{r}}{\partial n}\,\mathrm{d}S \tag{3.15}$$

where $F = V_1 - V_3$ and $P_0 \in CD^-$.

From the double layer property, we find at the boundary S that

$$F(P) = \frac{1}{4\pi}\frac{\sigma_1 - \sigma_2}{\sigma_1}\int_S F(Q)\frac{\partial \frac{1}{r}}{\partial n}\,\mathrm{d}S + \frac{1}{2}\frac{\sigma_1 - \sigma_2}{\sigma_1}F(P), \quad r = PQ. \tag{3.16}$$

After simple transformations, (3.16) takes the form

$$F(P) = \frac{1}{2\pi}k\int_S F(Q)\frac{\partial \frac{1}{r}}{\partial n}\,\mathrm{d}S, \quad k = \frac{\sigma_1 - \sigma_2}{\sigma_1 + \sigma_2}. \tag{3.17}$$

Thus, the function F is a solution to the characteristic equation

$$kAF = F$$

where A is the operator of the right-hand side of (3.17). The least characteristic value of the operator A is knoty, whereas, at finite positive values of σ_1 and σ_2, the inequality $|k| < 1$ is true. Therefore, equation (3.17) has only the trivial solution $F = 0$, i.e., everywhere outside D, $V_1 = V_3$. Thus, the following theorem holds.

Theorem 1.6. *For a function V_1 harmonic outside D, to be the electric potential of this domain, induced by a field W of sources located outside D, it is necessary and sufficient that V_1, be represented in the form (3.10).*

Note that the necessity was proven before.

As was noted above, problems of electric and magnetic prospecting are similar, and the author had obtained the expression for the magnetic potential of the domain D (in this case, σ_1 and σ_2 are magnetic permeabilities)

$$\mathbf{V}_1(P_0) = \frac{1}{4\pi}\frac{\sigma_1 - \sigma_2}{\sigma_2}\int\limits_S \left[\frac{\partial W(P)}{\partial n}\frac{1}{r} + \frac{\partial V_1(P)}{\partial n}\frac{1}{r}\right] \mathrm{d}S.$$

Repeating the above consideration for a simple layer, one can easily check that (3.11) is the necessary and sufficient condition for V_1 to be the magnetic potential of the domain D (in the framework of the problem statement).

From (3.6), taking into account the boundary conditions, we easily find

$$\mathbf{V}_2(P_0) = \frac{1}{4\pi}\frac{\sigma_1 - \sigma_2}{\sigma_2}\int\limits_S \left[\frac{\partial W(P)}{\partial n}\frac{1}{r} + V_2(P)\frac{\partial \frac{1}{r}}{\partial n}\right] \mathrm{d}S. \tag{3.18}$$

A similar theorem is valid for the function V_2 and relation (3.18).

3.4. THE ALGORITHM OF SOLUTION OF (3.10) AND (3.14) IN THE CLASS OF STELLAR BODIES

Bodies stellar in relation to a certain internal point form a broad class, important in practice. We will construct a spherical system of coordinates with the center at that point. Performing in (3.10) and (3.14) a spherical change of variables, we obtain

$$\mathbf{V}_1(P_0) = \frac{1}{4\pi}\frac{\sigma_1 - \sigma_2}{\sigma_1} \int_0^{2\pi}\int_0^{\pi} U(\rho(\theta,\varphi);\theta,\varphi)\frac{(\mathbf{r}^0 - \rho, \mathbf{N})}{|\,\mathbf{r}^0 - \rho\,|^3}\,\mathrm{d}\theta\,\mathrm{d}\varphi, \tag{3.19}$$

$$\mathbf{r}^0 = (x^0, y^0, z^0),\ \mathbf{N} = [\rho_\theta \times \rho_\varphi],\ U(\rho(\theta,\varphi);\theta,\varphi) = \text{const}$$

i.e.,

$$U(\rho(\theta,\varphi),\theta,\varphi) = U(\rho(\theta_{k^0},\varphi_{k^0});\theta_{k^0},\varphi_{k^0}) \tag{3.20}$$

(the choice of the point $(\theta_{k^0},\varphi_{k^0})$ is arbitrary), where $\rho(\theta,\varphi)$ is the right-hand side of the equation of the body surface $\rho = \rho(\theta,\varphi)$ in spherical coordinates.

Relations (3.19) and (3.20) may be considered as integro-differential equation and functional equation relative to the function $\rho(\theta,\varphi)$.

The right-hand side of the equation of the surface of the body D may be represented by a double Fourier series:

$$\rho(\theta,\varphi) = \sum_{k=-\infty}^{\infty}\sum_{j=-\infty}^{\infty} a_{kj}e^{i(k\theta+j\varphi)}. \tag{3.21}$$

We will therefore seek the solutions of (3.19) and (3.20) in the form of a segment of the double Fourier series

$$\rho_{nm}(\theta,\varphi) = \sum_{k=-n}^{n}\sum_{j=-m}^{m} a_{kj}e^{i(k\theta+j\varphi)} \tag{3.22}$$

minimizing the respective functionals

$$f(\gamma) = \sum_i^M [\mathbf{V}_1(x_i,y_i,z_i) - \mathbf{V}_1^{nm}(x_i,y_i,z_i)]^2 \tag{3.23}$$

$$g(\gamma) = \sum_j \{U[\rho_{nm}(\theta_j,\varphi_j),\theta_j,\varphi_j] - U[\rho_{nm}(\theta_{k^0},\varphi_{k^0});\theta_{k^0},\varphi_{k^0}]\}^2 \tag{3.24}$$

where the point (x_i,y_i,z_i) belongs to the information carrier which in the case of TIP may be chosen; γ is the vector of the coefficients of the function

$\rho_{nm}(\theta,\varphi)$, for which minimization is performed; and $\mathbf{V}_1^{nm}$ is the right-hand side of (3.19) substituting ρ by ρ_{nm}.

As was noted in (Tsirul'skii and Prutkin, 1981), the function $\rho(\theta,\varphi)$, which defines the boundary of the one-connected three-dimensional domain, must have the following property:

$$\rho(0,\varphi) = \text{const}, \qquad \rho(\pi,\varphi) = \text{const}. \tag{3.25}$$

From conditions (3.25), the relations for the coefficients ρ_{nm} follow:

$$\begin{gathered}\sum_k a_{kj} = 0, \quad \sum_{k_1} a_{k_1 j} = 0, \quad j \in [-m,m]; \\ k, k_1 \in [-n,n], \qquad k = \pm 1, \pm 3 \ldots; \quad k_1 = \pm 2, \pm 4, \ldots .\end{gathered} \tag{3.26}$$

We must thus find the minimum of the functional f and g subject to equality-type constraints. Practically, we add to the functionals f and g the functional

$$f_1 = \sum_j P_j \left(\left| \sum_k a_{kj} \right|^2 + \left| \sum_{k_1} a_{k_1 j} \right|^2 \right)$$

where P_j are the penalty coefficients. Normalization of (3.19) is done as in (Tsirul'skii and Prutkin, 1981) by using an analog of smoothing first-order Tikhonov normalizer

$$\Omega(\rho) = \int_0^{2\pi}\int_0^{\pi} (|\, \rho_\theta \,|^2 + |\, \rho_\varphi \,|^2) d\theta d\varphi = 2\pi^2 \sum_{k=-n}^{n} \sum_{j=-m}^{m} (k^2 + j^2) \,|\, a_{kj} \,|^2 . \tag{3.27}$$

In solving (3.15), the functional

$$f + f_1 + \alpha\Omega \tag{3.28}$$

was minimized, where the normalization parameter α was determined from the discrepancy. The Powell method was used for minimization (Tsirul'skii and Prutkin, 1981); the integral in (3.19) was computed by using the formulas of a high trigonometric accuracy, which in this case are formulas with equal weights and equidistant nodes (Krylov, 1959). In solving (3.20), a sphere of a minimum radius including all the singularities V_1 was taken for the zero approximation. The solution of (3.20) as the zero approximation was then defined for (3.19) (which includes the conductivity σ_2 as a parameter, making it possible to construct an entire family of ε-equivalent bodies generating with different conductivities the same field) for a certain value of $\sigma_2 = \sigma_0$. At each subsequent step the body D_{n-1} was taken as the zero approximation for the conductivity $\sigma_n < \sigma_{n-1}$. A sphere known *a priori* to contain the desired solution was chosen for the information carrier.

Examples of numeric realization. The algorithm, programmed in the Fortran language, have been tested in several theoretical examples.

Example 1. An ellipsoid of rotation with semiaxes 5:5:3 in a point source field was taken as the model. In this case, a representation of V_1 in the form of a series exists. For the potential of the ellipsoid, (3.20) was solved. Figure 3.1 shows the cross-sections cut off by the coordinate planes on ellipsoid (1) and on the solutions of (3.20).

Example 2. In this example, three point sources were taken for the model. An equivalent family of TIP was constructed: an 'ideal' conductor and solutions of (3.19) for two values of σ_2. Figure 3.2 shows the cross-sections cut off by the coordinate planes on the resulting bodies. The solution time for one value of the conductivity σ_2 was equal approximately 10 min.

3.5. ON THE SOLUTION OF THE CHARGE METHOD INVERSE PROBLEM

Source of primary field is located into abnormal body in the case of the charge method. Hence W is analytical function in D. From relations (3.9) we have

$$\mathbf{V}_1(P_0) = \frac{1}{4\pi}\frac{\sigma_1 - \sigma_2}{\sigma_1} \int\limits_S \left[\frac{\partial W(P)}{\partial n} + \frac{\partial V_2(P)}{\partial n}\right] \frac{\mathrm{d}S}{R}, \quad P_0 \in C\overline{D} \tag{3.29}$$

hence, by virtue of (3.2), we obtain

$$\mathbf{V}_1(P_0) = \frac{1}{4\pi}\frac{\sigma_1 - \sigma_2}{\sigma_1} \int\limits_S \left[\frac{\partial W(P)}{\partial n}\frac{1}{r} + V_1(P)\frac{\partial(1/r)}{\partial n}\right] \mathrm{d}S. \tag{3.30}$$

Similarly, substituting (3.7) into (3.6), we obtain

$$\mathbf{V}_2(P_0) = \frac{1}{4\pi}\frac{\sigma_1 - \sigma_2}{\sigma_2} \int\limits_S \left[\frac{\partial W(P)}{\partial n} + \frac{\partial V_1(P)}{\partial n}\right] \frac{\mathrm{d}S}{r}, \quad P_0 \in D \tag{3.31}$$

hence

$$\mathbf{V}_2(P_0) = \frac{1}{4\pi}\frac{\sigma_1 - \sigma_2}{\sigma_2} \int\limits_S [W(P) + V_1(P)] \frac{\partial(1/r)}{\partial n} \mathrm{d}S. \tag{3.32}$$

Relation (3.30) may be considered as the TIP equation for determining the boundary S. By stipulation of different σ_2 values it is possible to construct a single-parameter family ε of equivalent bodies (functions describing their boundary - solutions of equation (3.30)). Then, using the measured $\mathbf{V}_2$ values from equation (3.32) it is possible to determine σ_2. If the potential $\mathbf{V}_2$ was measured at a sufficiently large (in comparison with the number of parameters describing S) number of points; (3.32) also may be considered as the TIP equation. Finally, if σ_2 has been determined (on the basis of induction logging data), formula (3.32) can be used in checking the hypothesis on the boundary S (by computing the right-hand side measured at least at one point).

3.6. ALGORITHM FOR SOLVING EQUATIONS (3.30) AND (3.32) IN THE CLASS OF STELLAR BODIES

The class of bodies stellar relative to some point O is quite broad from the practical point of view. We shall construct a spherical coordinate system with its center at this point. Performing spherical change of the variables in (3.30) and (3.32), we obtain

$$\mathbf{V}_1(P_0) = \frac{1}{4\pi}\frac{\sigma_1 - \sigma_2}{\sigma_1} \int\limits_0^{2\pi}\int\limits_0^{\pi} \left[\frac{\operatorname{grad} W(P(\rho,\theta,\varphi)) \cdot \mathbf{N}}{|\,\rho - \mathbf{r}^0\,|^3} - \right.$$

$$\left. - \frac{V_1(P(\rho,\theta,\varphi))(\rho - \mathbf{r}^0, \mathbf{N})}{|\,\rho - \mathbf{r}^0\,|^3} \mathrm{d}\theta \mathrm{d}\varphi\right] \tag{3.33}$$

$$\mathbf{V}_2(O) = \frac{1}{4\pi}\frac{\sigma_1 - \sigma_2}{\sigma_2} \int\limits_0^{\pi}\int\limits_0^{2\pi} U(P(\rho,\theta,\varphi)) \sin\theta \mathrm{d}\theta \mathrm{d}\varphi \tag{3.34}$$

where $U = W + V_1$, $\rho(\theta,\varphi)$ is the right-hand side of the equation for the S surface in spherical coordinates $\rho = \rho(\theta,\varphi), \rho - \mathbf{r}^0 = \mathbf{R}$, $\mathbf{N} = \mathbf{r}_\theta \times \rho_\varphi$.

Relation (3.33) is an integro-differential equation and (3.34) is an integral equation relative to the function $\rho(\theta,\varphi)$. An algorithm for solution of such equations has been described above in detail ($\rho(\theta,\varphi)$ was selected in the form of a segment of a double Fourier series and the coefficients of the series were determined); we shall not discuss it. We note that in the derivation of (3.34) it was assumed that V_2 was measured at the stellar point; this made possible to derive a relation without the derivatives ρ_θ and ρ_φ.

Remark. The horizontal ground/air interface may be taken into account by the reflections method known in electric prospecting. For example, on the left-hand side of equation (3.33) in place of $\mathbf{V}_1(P_0)$ there will be $\mathbf{V}_1(P_0) + \mathbf{V}_1(P_0^*)$, where P_0^* is a point symmetric relative to the interface; if the measurements are made at this interface, $P_0 \equiv P_0^*$.

3.7. THE MAGNETIC FIELDS OF SPREAD CURRENTS INVERSE PROBLEM EQUATIONS

During the last years of geophysical exploration, the magnetic fields of spread currents; i.e. currents induced in the ground by grounded electrodes (Kormil'tsev *et al.*, 1975; Semenov *et al.*, 1977), have been used effectively. In the general case, the direct problem cannot be solved explicitly; therefore, the inverse problem is reduced to an operator equation of the first kind with an operator not defined explicitly.

However, such equations may be obtained for the theoretical inverse problem (TIP) of electric exploration, when from an anomalous electric or magnetic field given explicitly one has to find the bodies with various conductivities (magnetic permeabilities) equivalent in terms of the external field they generate (Martyshko, 1986, 1988). Note that the solution of the TIP is of interest not only as a realization of one stage in a two-stage interpretation method (Nikonova and Tsiryul'skii, 1978), but also it allows us to constructing geologically meaningful equivalents of various classes of singular sources. The aim of this chapter is to develop (on the basis of the explicit TIP equation for magnetic fields of conduction currents) a two-stage interpretation method of ground electric exploration data.

Mathematical interpretation problems of these fields are formulated in the following way.

3.7.1. Problem statement

Let a body T be a conducting inclusion with the parameters $\sigma_2 = \text{const}$, $\mu_2 = \text{const}$; it is placed in a medium with conductivity σ_1 and magnetic permeability μ_1. Assume also that there are direct current source points in the medium, and $\mathbf{H}$ and $\mathbf{E}$ are magnetic and electric fields, respectively (within and outside the conducting inclusion) created by these sources. Further we assume that T is a three-dimensional domain, S is its boundary, and $\mathbf{r} = \{x, y, z\}$ is the radius of a point from R^3.

As is known, the function vectors $\mathbf{H}(r)$ and $\mathbf{E}(r)$ satisfy the following set of equations (within and outside T, respectively):

$$\begin{gathered} rot\mathbf{E} = 0, \;\; \operatorname{div}\mathbf{E} = 0 \\ rot\mathbf{H} = j, \;\; \operatorname{div}\mathbf{H} = 0, \; \mathbf{j} = \sigma E \end{gathered} \tag{3.35}$$

and on the boundary S we have

$$\begin{gathered} [\mathbf{H}_2 - \mathbf{H}_1, \mathbf{n}] = 0, \;\; (\mu_2\mathbf{H}_2 - \mu_1\mathbf{H}_1, \mathbf{n}) = 0 \\ [\mathbf{E}_2 - \mathbf{E}_1, \mathbf{n}] = 0, \;\; (\sigma_1\mathbf{E}_1 - \sigma_2\mathbf{E}_2, \mathbf{n}) = 0. \end{gathered} \tag{3.36}$$

To solve a direct problem is to find $\mathbf{H}(r)$ and $\mathbf{E}(r)$, which must satisfy system (3.35) and boundary conditions (3.36). The inverse problem may be stated as follows: find the desired domain from a given function vector $\mathbf{H}(r)$, satisfying at the boundary of the domain conditions (3.36).

3.7.2. Equations of theoretical inverse problem

We shall call an inverse problem theoretical if $\mathbf{H}_1(r)$ is given explicitly (this situation arises, for example, after fitting the observed data to the field of singular sources). We have had an opportunity yet the implicit equations of inverse problems can be obtained if there is field integral representation of kind (3.4). For vector field analogy formulas exist as well. From conditions (3.35) and vector analog for Pompey formulas follows the representation:

$$\frac{1}{4\pi}\int\limits_S \Big\{(\mathbf{n}, \mathbf{H}^2_\alpha)\operatorname{grad}\frac{1}{|r-r'|} + \mathbf{n}\times\mathbf{H}^2_\alpha\times\operatorname{grad}\frac{1}{|r-r'|}$$

$$+\mathbf{n}\times rot\mathbf{H}^2_\alpha\frac{1}{|r-r'|}\Big\}\,dS = \begin{cases} H^2_\alpha(r'), & r'\in D, \\ 0, & r'\in C\overline{D} \end{cases} \tag{3.37}$$

$$\frac{1}{4\pi}\int_S \left\{(\mathbf{n}, \mathbf{H}^1_\alpha) \operatorname{grad} \frac{1}{|r - r'|} + \mathbf{n} \times \mathbf{H}^1_\alpha \times \operatorname{grad} \frac{1}{|r - r'|}\right.$$

$$\left. + \mathbf{n} \times rot\mathbf{H}^1_\alpha \frac{1}{|r - r'|}\right\} \mathrm{d}S = \begin{cases} -H^1_\alpha(r'), & r' \in C\overline{D}, \\ 0, & r' \in D. \end{cases} \qquad (3.38)$$

Subtracting from (3.37) equation (3.38) when $r \in C\overline{D}$ and in view that $rot\ \mathbf{H} = \sigma\mathbf{E}$ we obtain

$$\mathbf{H}^1_\alpha(r') = \frac{1}{4\pi}\int_S \left\{(\mathbf{n}, \mathbf{H}^2_\alpha - \mathbf{H}^1_\alpha) \operatorname{grad} \frac{1}{|r - r'|}\right.$$

$$\left. + \mathbf{n} \times \left(\mathbf{H}^2_\alpha - \mathbf{H}^1_\alpha\right) \times \operatorname{grad} \frac{1}{|r - r'|} + \mathbf{n} \times \left(\sigma_2 \mathbf{E}^2_\alpha - \sigma_1 \mathbf{E}^1_\alpha\right) \frac{1}{|r - r'|}\right\} \mathrm{d}S. \qquad (3.39)$$

From formula (3.39) and boundary conditions (3.35) it follows that

$$\mathbf{H}^1_\alpha(r') = \frac{1}{4\pi}\left\{\frac{\mu_1 - \mu_2}{\mu_2}\int_S (\mathbf{n}, \mathbf{H}^1) \operatorname{grad} \frac{1}{|r - r'|} \mathrm{d}S\right.$$

$$\left. + (\sigma_2 - \sigma_1)\int_S \mathbf{n} \times \mathbf{E}^1_\alpha \frac{\mathrm{d}S}{|r - r'|}\right\}. \qquad (3.40)$$

Taking into account $\mathbf{E}^1 = \frac{1}{\sigma_1} rot\mathbf{H}^1$, we obtain

$$\mathbf{H}^1_\alpha(r') = \frac{1}{4\pi}\int_S \left\{(\mathbf{n}, \frac{\mu_1 - \mu_2}{\mu_2}\mathbf{H}^1) \operatorname{grad} \frac{1}{|r - r'|}\right.$$

$$\left. + \frac{\sigma_2 - \sigma_1}{\sigma_1}\mathbf{n} \times rot\mathbf{H}^1_\alpha \frac{1}{|r - r'|} \mathrm{d}S\right\}. \qquad (3.41)$$

An induced magnetic field arising as a result of medium inhomogeneities in weak magnetic rocks (i.e., $\mu_2 \approx \mu_1 = 1$) is considerably less than the magnetic field of currents and in most cases it may be neglected. Hence at $\mu_2 \approx \mu_1$, instead of (3.41) for solution of the inverse problem, the following equation may be used

$$\mathbf{H}^1_\alpha(r') = \frac{1}{4\pi}\frac{\sigma_2 - \sigma_1}{\sigma_1}\int\limits_S \mathbf{n}\times rot\mathbf{H}^1_\alpha \frac{dS}{|\, r - r'\,|} \tag{3.42}$$

and instead of (3.40) one may use

$$\mathbf{H}^1_\alpha(r') = \frac{1}{4\pi}(\sigma_2 - \sigma_1)\int\limits_S \mathbf{n}\times \mathbf{E}^1_\alpha \frac{dS}{|\, r - r'\,|}. \tag{3.43}$$

Therefore,

$$\mathbf{H}^1_\alpha(r') = \frac{1}{4\pi}(\sigma_2 - \sigma_1)\int\limits_S \mathbf{n}\times \operatorname{grad} \mathbf{V}_1 \frac{dS}{|\, r - r'\,|} \tag{3.44}$$

is a connection equation between the electric field given on the surface S and measured values of H.

Analogous representations hold for an internal field, when $r' \in T$:

$$\mathbf{H}^2_\alpha(r') = \frac{1}{4\pi}\int\limits_S \Big\{ (\mathbf{n}, \frac{\mu_1 - \mu_2}{\mu_2}\mathbf{H}^1) \operatorname{grad}\frac{1}{|\, r - r'\,|}$$

$$+ \frac{\sigma_2 - \sigma_1}{\sigma_1}\mathbf{n}\times rot\mathbf{H}^1_\alpha \frac{1}{|\, r - r'\,|}\Big\} dS$$

and

$$\mathbf{H}^2_\alpha(r') = \frac{1}{4\pi}\frac{\sigma_2 - \sigma_1}{\sigma_1}\int\limits_S \mathbf{n}\times rot\mathbf{H}^1_\alpha \frac{dS}{|\, r - r'\,|}.$$

To solve (3.40)–(3.44), we may use the algorithm developed and programmed for a computer for the class of stellar bodies.

3.8. TWO-STAGE METHOD OF THE SOLUTION TO THE INVERSE PROBLEM OF ELECTRIC EXPLORATION

As was noted in Chapter 1, on realizing the first stage (i.e. the approximation of observed data by fields of singular sources), it is extremely important to select an optimal class of these sources. In this class it should be possible to approximate (with good accuracy) the fields of the sufficiently wide (from the practical point of view) set of anomalous objects, with a minimal number of parameters being used. The attempt to carry out the complete theoretical study (as it was done by Nikonova and Tsirul'skii (1978) for two-dimensional gravimetric problems) fails in the three-dimensional case.

The results of mathematical simulation allow conclusions on the suitability of any type of singular sources for fitting the fields of the particular models. So, for example, Petrov and Fedorov (1988, 1989) used potentials of dipole sources for electric field approximation. The author programmed for a computer two-stage methods for solving the inverse problem for electric and magnetic fields.

We shall illustrate the application of the two-stage method using a model example.

Example 1. Electric field of inverse problem numerical result.

In order to construct the electric field we have used the sum of potentials for point sources and linear currents. 'Linear currents' is meant continuous set of point sources. Suppose, along a line l a charge q is distributed; points $M_1(x_1, y_1, z_1)$ and $M_2(x_2, y_2, z_2)$ are the end of the line. We have equations along M_1M_2 :

$$\begin{cases} x = x_1 + (x_2 - x_1)t, \\ y = y_1 + (y_2 - y_1)t, \\ z = z_1 + (z_2 - z_1)t, \ 0 \le t \le 1. \end{cases} \tag{3.45}$$

Then the electric potential V_1 can be calculated as follows

$$V_1(M_0) = \int_{M_1}^{M_2} \frac{q dl}{R} = \int_0^1 \frac{q l dt}{\sqrt{(x_0 - x)^2 + (y_0 - y)^2 + (z_0 - z)^2}} \tag{3.46}$$

From (3.42), we obtain

$$V_1(x_0, y_0, z_0) = \begin{cases} q \ln \left[\dfrac{l^2 - \mathbf{l} \cdot \mathbf{R}_{01} + l \cdot R_{02}}{l \cdot R_{01} - \mathbf{l} \cdot \mathbf{R}_{01}} \right] \\ \\ q \ln \left[\frac{R_{01} - l}{R_{01}} \right], \text{ if } \mathbf{l} \| \mathbf{R}_{01} \end{cases} \tag{3.47}$$

where

$$\mathbf{R}_{01} = \{x_0 - x_1, y_0 - y_1, z_0 - z_1\}, R_{01} = |\mathbf{R}_{01}|,$$

$$\mathbf{R}_{02} = \{x_0 - x_2, y_0 - y_2, z_0 - z_2\}, R_{02} = |\mathbf{R}_{02}|,$$

$$\mathbf{l} = \{x_2 - x_1, y_2 - y_1, z_2 - z_1\}, l = |\mathbf{l}|.$$

For ellipsoid with $a = 5$, $b = 5$, $c = 1$ we have solved the direct problem; the data of potential V_1 was calculated in the plane $z = 0$. Then we have approximated V_1 with potentials of two point sources and linear currents and solved the inverse problem (i.e. (3.19)). We see that the results are good enough.

Example 2. Consider the example of solution of the inverse problem for the magnetic field.

In the case of the inverse problem for the magnetic field, the magnetic field of linear currents were used for fitting. As is known (Gurevich, 1982), the magnetic fields of spread currents may be approximated well enough by the fields of equivalent linear currents in many cases which are of a practical interest.

By linear current we shall mean a current flowing in a conductor with the thickness negligible in comparison with the distance from it to the observation point. Accoding with the Beeo-Savar formula

$$\mathbf{H} = \frac{1}{4\pi} \int\limits_V \frac{\mathbf{R} \times \mathbf{j}}{R^3} \mathrm{d}V \tag{3.48}$$

where $\mathbf{R}$ is the radius vector from observation point to current element $\mathbf{j} \cdot \mathrm{d}V$; so, for elementary current, we have

$$\mathrm{d}\mathbf{H} = \frac{1}{4\pi} \frac{\mathbf{R} \times \mathbf{j}}{R^3} \mathrm{d}V. \tag{3.49}$$

Let $\mathrm{d}V = \mathrm{d}\mathbf{l} \cdot \mathrm{d}s$, where $\mathrm{d}s$ is the square of current section, $d\mathbf{l}$ is the length element. Then we get

$$\mathbf{j} \cdot \mathrm{d}V = \mathbf{j} \cdot \mathrm{d}s \cdot \mathrm{d}\mathbf{l} = J \cdot \mathrm{d}\mathbf{l}. \tag{3.50}$$

Accoding with (3.49) and (3.50) we have

$$\mathrm{d}\mathbf{H} = \frac{J}{4\pi} \frac{\mathbf{R} \times \mathrm{d}\mathbf{l}}{R^3} \mathrm{d}V \tag{3.51}$$

which holds for arbitrary linear current systems.

As far as a wire of any configuration may be approximated by a broken line, the case of the linear current is of particular interest. Let J be a current flowing through the linear wire of length $2L$ from the point d to the point c. We guide the axis z of the cylindrical coordinate system (z, r, φ) along the wire in the direction of current flow and choose the coordinate origin in the middle of the wire. Choosing the coordinate system in this way, we have only the azimuthal component. The total magnetic field of the wire length therefore is calculated as a sum taken over elementary strengths

$$H_\varphi = \int_c^d dH_\varphi = -\frac{Jr}{4\pi} \int_{-L}^{L} \frac{dl}{R^3}. \tag{3.52}$$

Whence, for the calculation of the magnetic field of a current-carrying wire, we obtain

$$H_\varphi = \frac{J}{4\pi r}\left[\frac{z+L}{\sqrt{r^2+(z+L)^2}} - \frac{z-L}{\sqrt{r^2+(z-L)^2}}\right]. \tag{3.53}$$

In the general case, in the Cartesian coordinate system, components of the magnetic field of linear current H_x, H_y, H_z are found as projections of H_φ on the corresponding axis (with respect to the linear current center position).

Thus, the magnetic field of linear current is described by seven parameters and may be used to approximate the fields of a parallelepiped (see Gurevich, 1982).

The direct problem was solved for the parallelepiped with the ratio of the edge lengths $L_x : L_y : L_z = 10 : 5 : 1$ to the center at the depth of 5 ($\sigma_2 = 100$, $\sigma_1 = 1$). Then, calculated in the plane $z = 0$, the values of H were fitted by the field of linear currents (see Figure 3.1). For the field of these sources, the family of solutions of the inverse problem was constructed (a one-parameter family of solutions to(5)). As we see from Figure 3.1, the parallelepiped is reconstructed well enough (to within the algorithm error).

We shall illustrate now opportunity of effective allocation of the unique solution of the inverse problem on data of supervision of an electric field in two cases of source position of a normal field. In Figure 3.5, two equivalent families of the solutions (a_1 and a_2 are the two positions of sources of a normal field) are constructed; 1 is the equivalent family of Example 2, Section 3.3. From one of the terms of the family with $\sigma = 20$ by the algorithm, developed above (Sections 2.2–2.3), (the direct problem in a normal field of a source a_2) the values of electric potential V_1 were calculated. Further, by the method described above, we have constructed the equivalent family of the solutions of the inverse problem for V_1. As we see from Figure 3.5, the areas differs minimally for $\sigma = 20$.

Conclusion. The basic results of Chapter 3: for the TIP solution of the electric exploration on a constant current and for the method of artificial field magnetization we have obtained the equation with the explicitly given operator simpler than those known before. Numerical algorithms for solution of equations (3.19) and (3.20) were developed. The examples of equivalent families of the TIP solutions of electric exploration were constructed. Together with any trial and error method of selection of three-dimensional fields by singular sources on this basis, it is possible to construct methods of interpretation of the given electric exploration, taking into account the equivalent solutions.

For the solution of the theoretical inverse problem by the charge method using measurements of internal and external electric potential of conducting inclusion, the equations with the explicitly given operator were obtained. The joint solution of these equations allows us to determine the boundary and conductivity of the unknown object.

Explicit TIP equations for magnetic field of spreading currents are obtained, i.e. basic transition from scalar to vector field was made. Two-stage algorithm of the solution of the inverse problem of electric exploration was realized; first examples of the equivalent solutions of the three-dimensional inverse problem were constructed.

We shall emphasize in summary that Petrov (Petrov and Fedorov, 1988; Sobolev, 1962) on the basis of algorithms of solution of the equation (3.20), developed by the author, have constructed practical examples. Thus, at a stage of selection, the potentials of dipoles were used.

Apparently, in the three-dimensional case it is not possible to determine an optimal class of functions for selection of observed data (there are no classes of solvability in finite form). The equations of theoretical inverse problems are universal; only they allow us to solve TIP for any class of potentials.

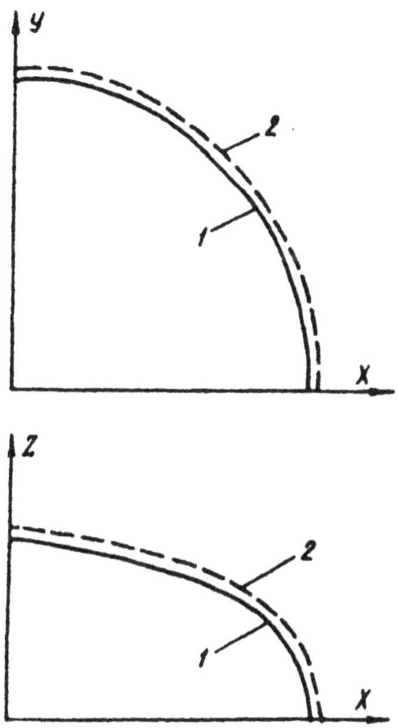

Figure 3.1

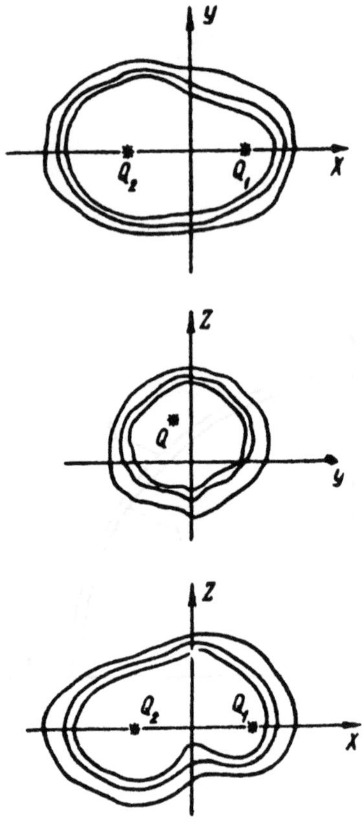

Figure 3.2

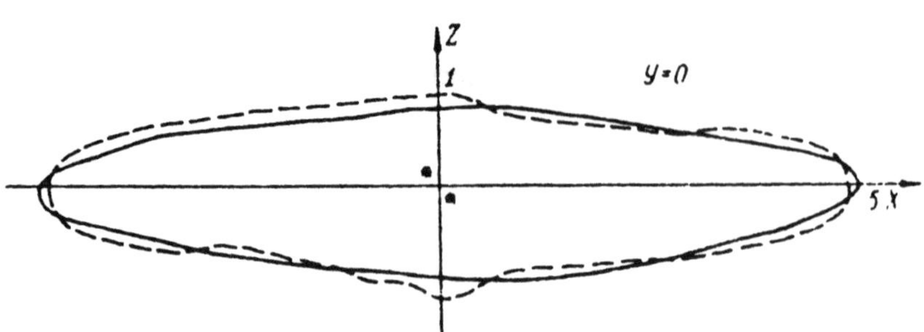

Figure 3.3

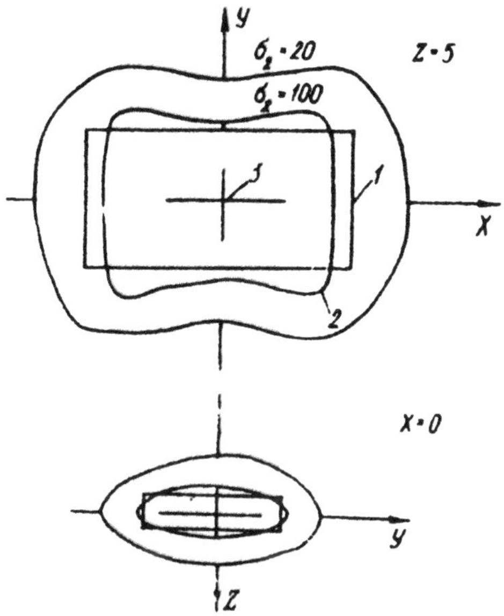

Figure 3.4

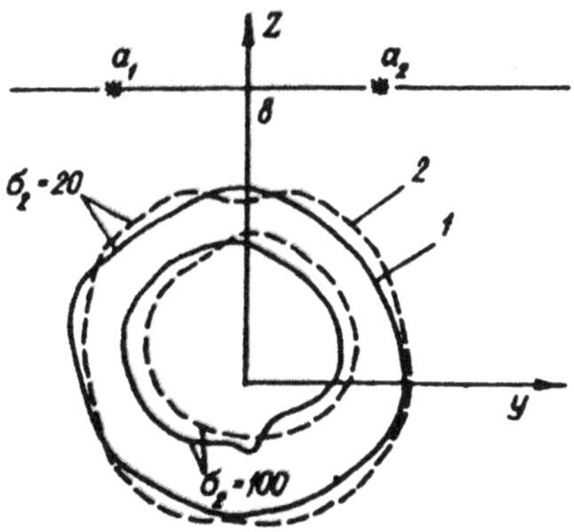

Figure 3.5

Chapter 4.

Explicit equations for inverse problems of electromagnetic field

Explicit integro-differential equations of theoretical inverse problems satisfying the Helmholtz, telegrapher's, and diffusion equations are derived. We have used representations of field functions through the values of the functions themselves and the derivatives at the boundary of the anomaly-forming object. The primary field source may be of an arbitrary type. As a result of interpretation we obtain the bodies stellar relative to some point with different values of conductivity which generated the same (electric or magnetic) field. Functional equations for a "perfect conductor" are presented.

Electromagnetic methods are now widely used in geophysical prospecting. The interpretation of data obtained by these methods involves considerable difficulties because in a general case the inverse problem is reduced to an operator equation of the first kind with an implicitly stipulated operator. The numerical solution of such equations requires considerable expenditures of computer time. However, in the case of the theoretical inverse problem (TIP) in electric prospecting with a direct current (dc), the author was able to obtain explicit integro-differential equations for the electric and magnetic fields (Martyshko, 1986 a, 1988, 1990 a, b), to develop effective algorithms for solving these equations, and to construct examples of equivalent domains. We note that solution of the TIP not only is of an interest as a realization of the stage in the two-stages interpretation method but also makes it possible to construct geologically meaningful equivalents for different classes of singular sources. The first interpretation stage is approximating the observed data with the fields of singular sources. There are no significant theoretical problems on this way, so we paid attention to the second stage. It seems that we have not only to solve the TIP equations, but also the methods for deriving them; merit attention: representations of field functions through the values of the functions themselves and the derivatives at the boundary of the anomaly-forming object are used (in Martyshko (1986 a), the main formula of the theory of harmonic functions; in Martyshko (1988, 1989), the Stratton-Chu formulas (Stratton, 1948)). The main goal of this chapter is to derive TIP equations for different types of variable electromagnetic fields by the method developed earlier. We shall not limit ourselves to a quasi-stationary monochromatic field (that is, the Helmholtz equation) because in some geophysical methods (such as MPP) it is necessary to deal with telegrapher's equation.

4.1. EQUATIONS OF ELECTRODYNAMICS

In the theory of geophysical field we deal with the Helmholtz equations

$$rot\mathbf{H} = \mathbf{j} + \mathbf{j}^e + \frac{\partial \mathbf{D}}{\partial t} \tag{4.1}$$

$$rot\mathbf{E} = -\frac{\partial \mathbf{B}}{\partial t}, \tag{4.2}$$

$$\operatorname{div}\mathbf{B} = 0 \tag{4.3}$$

$$\operatorname{div}\mathbf{D} = q + q^e, \tag{4.4}$$

where $\mathbf{E}$ and $\mathbf{D}$ are vectors of electric field; $\mathbf{H}$ and $\mathbf{B}$ characterize magnetic field; $\mathbf{j}$ is the density of conductivity currents; q is the volume density of free charges; $\mathbf{j}^e$, q^e are the densities of external currents and charges. In this case,

$$\operatorname{div}\mathbf{j}^e = -\partial q^e/\partial t. \tag{4.5}$$

We shall suppose that the medium is linear and isotropic

$$\mathbf{D} = \varepsilon\mathbf{E}, \qquad \mathbf{B} = \mu\mathbf{H} \tag{4.6}$$

where ε, μ are dielectric permittivity and magnetic permeability respectively;

$$\mathbf{j} = \sigma \mathbf{E} \tag{4.7}$$

where σ is the conductivity.

If ε, μ, σ are constants, we have

$$rot\mathbf{H} = \sigma\mathbf{E} + \mathbf{j}^e + \varepsilon\partial\mathbf{E}/\partial t \tag{4.8}$$

$$rot\mathbf{E} = -\mu\partial\mathbf{H}/\partial t \tag{4.9}$$

$$\operatorname{div}\mathbf{H} = 0 \tag{4.10}$$

$$\operatorname{div}\mathbf{E} = (q + q^e)/\varepsilon. \tag{4.11}$$

We may obtain the separate equations for **E** and **H**

$$\Delta\mathbf{E} - \mu\varepsilon\partial^2\mathbf{E}/\partial t^2 - \mu\sigma\partial\mathbf{E}/\partial t = \mu\partial\mathbf{j}^e/\partial t + \frac{1}{\varepsilon}\operatorname{grad}(q + q^e) \tag{4.12}$$

$$\Delta\mathbf{H} - \mu\varepsilon\partial^2\mathbf{H}/\partial t^2 - \mu\sigma\partial\mathbf{H}/\partial t = -rot\mathbf{j}^e. \tag{4.13}$$

If there are no external currents and charges, we have

$$\Delta\mathbf{E} - \mu\varepsilon\partial^2\mathbf{E}/\partial t^2 - \mu\sigma\partial\mathbf{E}/\partial t = 0 \tag{4.14}$$

$$\Delta\mathbf{H} - \mu\varepsilon\partial^2\mathbf{H}/\partial t^2 - \mu\sigma\partial\mathbf{H}/\partial t = 0. \tag{4.15}$$

Equations (4.14), (4.15) are the so-called telegrapher's equations.

4.2. FORMULATION OF PROBLEM

Assume that in a linear isotropic medium with conductivity σ_1 and permeability μ_1, there is an inclusion, a body T with parameters σ_2, μ_2. Also assume that in the medium, there are sources of an electromagnetic field and that $\mathbf{H}_1, \mathbf{E}_1, \mathbf{H}_2, \mathbf{E}_2$ are the strengths of the magnetic and electric fields outside and inside the conducting inclusion, respectively, caused by these sources. Henceforth we shall assume that T is a three-dimensional domain, S is its boundary, $r = \{x, y, z\}$ is the radius vector of a point from R^3.

As is well known, in a linear isotropic medium (σ, ε, μ are constants) in the absence of stray currents and charges the function vectors $\mathbf{H}, \mathbf{E}$ satisfy the following system of equations (inside and outside T, respectively):

$$rot\mathbf{H} = \mathbf{j} + \frac{\partial \mathbf{D}}{\partial t}, \tag{4.1'}$$

$$rot\mathbf{E} = -\frac{\partial \mathbf{B}}{\partial t}, \tag{4.2'}$$

$$\operatorname{div} \mathbf{B} = 0, \tag{4.3'}$$

$$\operatorname{div} \mathbf{D} = 0, \tag{4.4'}$$

where $\mathbf{D} = \varepsilon \mathbf{E}$, $\mathbf{B} = \mu \mathbf{H}$, $\mathbf{j} = \sigma \mathbf{E}$ is the density of conductivity currents.

In the boundary S, we have

$$[\mathbf{H}_2 - \mathbf{H}_1, \mathbf{n}] = 0, \;\; (\mu_2 \mathbf{H}_2 - \mu_1 \mathbf{H}_1, \mathbf{n}) = 0$$

$$[\mathbf{E}_2 - \mathbf{E}_1, \mathbf{n}] = 0, \;\; (\varepsilon_2 \mathbf{E}_2 - \varepsilon_1 \mathbf{E}_1, \mathbf{n}) = \eta \tag{4.16}$$

ε is a permittivity; η is a surface density of the electric charge. Assume that $\mathbf{E}_1^\alpha, \mathbf{H}_1^\alpha$ is the external field of the domain T induced by the sources of primary field $\mathbf{H}^H$ and $\mathbf{E}^H$. Then

$$\mathbf{H}_1 = \mathbf{H}_1^\alpha + \mathbf{H}^H, \mathbf{E}_1 = \mathbf{E}_1^\alpha + \mathbf{E}^H \tag{4.17}$$

$$[\mathbf{H}^H + \mathbf{H}_1^\alpha, \mathbf{n}] = [\mathbf{H}_2, \mathbf{n}], \quad [\mathbf{E}^H + \mathbf{E}_1^\alpha, \mathbf{n}] = [\mathbf{E}_2, \mathbf{n}]$$

$$(\mathbf{H}_2, \mathbf{n}) = \frac{\mu_1}{\mu_2}(\mathbf{H}_1^\alpha + \mathbf{H}^H, \mathbf{n}), \quad (\mathbf{E}_2, \mathbf{n}) = [\varepsilon_1(\mathbf{E}_1, \mathbf{n}) + \eta]/\varepsilon_2. \tag{4.18}$$

The vectors of the electric and magnetic fields in the domain of medium homogeneity in the absence of stray currents and charges satisfy the homogeneous telegrapher's equation

$$\Delta\mathbf{F} - \mu\varepsilon\frac{\partial^2\mathbf{F}}{\partial t^2} - \mu\sigma\frac{\partial\mathbf{F}}{\partial t} = 0. \tag{4.19}$$

Thus, $\mathbf{H}_1^\alpha$ and $\mathbf{E}_1^\alpha$ satisfy (4.19) when $\sigma = \sigma_1, \mu = \mu_1, \varepsilon = \varepsilon_1$ outside the domain T; and $\mathbf{H}_2$ and $\mathbf{E}_2$ inside the domain T ($\sigma = \sigma_2, \mu = \mu_2, \varepsilon = \varepsilon_2$).

For a monochromatic field, it follows from (4.8)-(4.11) that:

$$rot\mathbf{H} = \sigma^*\mathbf{E}, \; rot\mathbf{E} = i\omega\mu\mathbf{H}, \; \operatorname{div}\mathbf{H} = 0, \; \operatorname{div}\mathbf{E} = 0 \tag{4.20}$$

where $\sigma^* = \sigma - i\omega\varepsilon$ is the complex conductivity; in this case (4.19) is transformed into the Helmholtz equation

$$\Delta\mathbf{F} + k^{*2}\mathbf{F} = 0 \tag{4.21}$$

where the wave number of the medium $k^* : k^{*2} = i\omega\mu\sigma + \omega^2\mu\varepsilon$.

For a quasi-stationarity field, we have

$$rot\mathbf{H} = \sigma\mathbf{E}, \quad rot\mathbf{E} = -\mu\frac{\partial\mathbf{H}}{\partial t} \tag{4.22}$$

and the vectors $\mathbf{E}$ and $\mathbf{H}$ satisfy the diffusion equation

$$\Delta\mathbf{F} - \mu\sigma\frac{\partial\mathbf{F}}{\partial t} = 0. \tag{4.23}$$

Let $\sigma = 0$, then (4.19) is transformed to the equation

$$\Delta\mathbf{F} - \mu\sigma\frac{\partial^2\mathbf{F}}{\partial t^2} = 0 \tag{4.24}$$

and equation (4.23) is reduced to

$$\Delta\mathbf{F} = 0. \tag{4.25}$$

Both equations are satisfied by functions $\mathbf{H}_1$ and $\mathbf{E}_1$ inside the domain T.

Definition 4.1. We formulate the following inverse problem: to determine a domain T on the given function vector $\mathbf{F}(r)$ satisfying (4.19), (4.21), (4.23), (4.24) outside the domain T and satisfying conditions (4.18) at the boundary of the unknown domain S ($\mathbf{H}$ and $\mathbf{E}$ in (4.19), (4.23), (4.24) naturally must satisfy the relevant initial conditions).

Definition 4.2. Following Tsirul'skii (1976), an inverse problem is said to be the theoretical inverse problem, provided that $\mathbf{H}_1(\mathbf{E}_1)$ is given in an explicit form (a similar situation occurs, for example, upon describing observed data by selected fields of singular sources).

4.3. EXPLICIT TIP EQUATIONS FOR MONOCHROMATIC FIELD

Assume that $\mathbf{F}$ and P are arbitrary functions twice continuously differentiable in T (as far as in the boundary S). Then the following integral relation holds (Zhdanov, 1984):

$$\begin{aligned}&\int_T (\Delta P\cdot\mathbf{F}+\operatorname{grad}P\operatorname{div}\mathbf{F}+P rot rot\mathbf{F})\mathrm{d}v\\&=\int_S\{(\mathbf{nF})\operatorname{grad}P+[\mathbf{n}\times\mathbf{F}]\times\operatorname{grad}P+[\mathbf{n}\times rot\mathbf{F}]P\}\,\mathrm{d}s.\end{aligned}\tag{4.26}$$

Formula (4.26) plays an exceptionally important role in the theory of geophysical fields because for any field (not only electromagnetic!), it makes possible to obtain a representation through its boundary values: it is sufficient to take a fundamental solution of the equation which satisfies $\mathbf{F}$ and substitute into (4.20). Such representations, in turn, can be used for deriving TIP equations.

The Stratton-Chu formulas can be derived for the monochromatic field from (4.26)

$$\begin{aligned}&\int_S\{(\mathbf{n},\mathbf{E}_2)\operatorname{grad}G_2+[\mathbf{n}\times\mathbf{E}_2]\times\operatorname{grad}G_2+i\omega\mu_2[\mathbf{n}\times\mathbf{H}_2]G_2\}\,\mathrm{d}s\\&\qquad=\begin{cases}\mathbf{E}_2(r'),\ r'\in T\\ \quad 0,\ r'\in C\bar{T}\end{cases}\end{aligned}\tag{4.27}$$

$$\int_S \{(\mathbf{n}, \mathbf{E}_1^\alpha) \operatorname{grad} G_1 + [\mathbf{n} \times \mathbf{E}_1^\alpha] \times \operatorname{grad} G_1 \ + i\omega\mu_1 [\mathbf{n} \times \mathbf{H}_1^\alpha] G_1\} \, \mathrm{d}s$$

$$= \begin{cases} 0, r' \in T \\ -\mathbf{E}_1^\alpha(r'), r' \in C\bar{T} \end{cases} \tag{4.28}$$

$$\int_S \{(\mathbf{n}, \mathbf{H}_2) \operatorname{grad} G_2 + [\mathbf{n} \times \mathbf{H}_2] \times \operatorname{grad} G_2 \ + \sigma_2^* [\mathbf{n} \times \mathbf{E}_2] G_2\} \, \mathrm{d}s$$

$$= \begin{cases} \mathbf{H}_2(r'), r' \in T \\ 0, r' \in C\bar{T} \end{cases} \tag{4.29}$$

$$\int_S \{(\mathbf{n}, \mathbf{H}_1^\alpha) \operatorname{grad} G_1 + [\mathbf{n} \times \mathbf{H}_1^\alpha] \times \operatorname{grad} G_1 \ + \sigma_1^* [\mathbf{n} \times \mathbf{E}_1^\alpha] G_1\} \mathrm{d}s$$

$$= \begin{cases} -\mathbf{H}_1^\alpha(r'), r' \in C\bar{T} \\ 0, r' \in T \end{cases} \tag{4.30}$$

$$G_{1,2}(r' \mid r) = -\frac{e^{ik_{1,2}^*|r-r'|}}{4\pi|r-r'|}. \tag{4.31}$$

Subtracting (4.28) from (4.27) when $r' \in C\bar{T}$ and using relations (4.18), (4.20), we obtain

$$\mathbf{E}_1^\alpha(r') = \int_S \Big\{ (\mathbf{n}, \mathbf{E}_1^\alpha)[\operatorname{grad}(\frac{\varepsilon_1}{\varepsilon_2} G_2 - G_1)] + \frac{\operatorname{grad} G_2}{\varepsilon_2}[\eta + (\mathbf{n}, \mathbf{E}^H)\varepsilon_1]$$

$$+ [\mathbf{n} \times \mathbf{E}_1^\alpha] \times \operatorname{grad}(G_2 - G_1) + [\mathbf{n} \times \mathbf{E}^H] \times \operatorname{grad} G_2$$

$$+ i\omega[\mathbf{n} \times \mathbf{H}_1^\alpha](\mu_2 G_2 - \mu_1 G_1) + i\omega\mu_2[\mathbf{n} \times \mathbf{H}^H] G_2 \Big\} \, \mathrm{d}s \tag{4.32}$$

where $\eta = \left(\frac{\varepsilon_1}{\sigma_1^*} - \frac{\varepsilon_2}{\sigma_2^*}\right) \operatorname{div}_S[\mathbf{n} \times \mathbf{H}_1]$, since $(\mathbf{n}, \mathbf{E})|_S = -\frac{1}{\sigma^*} \operatorname{div}_S[\mathbf{n} \times \mathbf{H}]$.

Similarly, from (4.29) and (4.30), we obtain

$$\mathbf{H}_1^\alpha(r') = \int_S \Big\{ (\mathbf{n}, \mathbf{H}_1^\alpha) \operatorname{grad}\left(\frac{\mu_1}{\mu_2} G_2 - G_1\right) + \frac{\mu_1}{\mu_2}(\mathbf{n}, \mathbf{H}^H) \operatorname{grad} G_2$$

$$+[\mathbf{n}\times\mathbf{H}_1^\alpha]\times\operatorname{grad}(G_2-G_1)+[\mathbf{n}\times\mathbf{H}^H]\times\operatorname{grad}G_2$$

$$+[\mathbf{n}\times\mathbf{E}_1^\alpha](\sigma_2^*G_2-\sigma_1^*G_1)+[\mathbf{n}\times\mathbf{E}^H]\sigma_2^*G_2\}\,\mathrm{d}s. \tag{4.33}$$

Relations (4.32) and (4.33) are the equations of the theoretical inverse problem for a monochromatic field (relative to the boundary S). The real properties of the anomalous domain are assumed to be taken as parameters, i.e., we construct the solution of the TIP for various values $\sigma_2, \varepsilon_2, \mu_2$. As the result, we obtain the equivalent family of bodies which generates the similar electric or magnetic field. In the case of a stationary field ($\omega=0$, $k^*=0$), the known equations of the inverse problem for dc electric prospecting are already derived from (4.32) and (4.33) (Martyshko, 1986 a, b; 1988):

$$\mathbf{E}_1^\alpha(r')=\frac{1}{4\pi}\frac{\sigma_2-\sigma_1}{\sigma_2}\int_S(\mathbf{n},\mathbf{E}_1^\alpha+\mathbf{E}^H)\operatorname{grad}\frac{1}{|r-r'|}\mathrm{d}s \tag{4.32'}$$

$$\mathbf{H}_1^\alpha(r')=\frac{1}{4\pi}\int_S\left\{\frac{\mu_1-\mu_2}{\mu_2}(\mathbf{n},\mathbf{H}_1)\operatorname{grad}\frac{1}{|r-r'|}+\frac{\sigma_2-\sigma_1}{\sigma_1}\frac{[\mathbf{n}\times rot\mathbf{H}_1^\alpha]}{|r-r'|}\right\}\mathrm{d}s. \tag{4.33'}$$

If the modified Stratton-Chu integrals (Henl *et al.*, 1964)

$$\frac{1}{\sigma_2^*}rot'rot'\int_S[\mathbf{n}\times\mathbf{H}_2]G_2\mathrm{d}S+rot'\int_S[\mathbf{n}\times\mathbf{E}_2]G_2\mathrm{d}S$$

$$=\begin{cases}\mathbf{E}_2(r'),\ r'\in T,\\ 0,\ r'\in C\bar{T}\end{cases} \tag{4.34}$$

$$\frac{1}{i\omega\mu_2}rot'rot'\int_S[\mathbf{n}\times\mathbf{E}_2]G_2\mathrm{d}S+rot'\int_S[\mathbf{n}\times\mathbf{H}_2]G_2\mathrm{d}S$$

$$=\begin{cases}\mathbf{H}_2(r'),\ r'\in T,\\ 0,\ r'\in C\bar{T}\end{cases} \tag{4.35}$$

$$\frac{1}{\sigma_2^*}rot'rot'\int_S[\mathbf{n}\times\mathbf{H}_1^\alpha]G_1\mathrm{d}S+rot'\int_S[\mathbf{n}\times\mathbf{E}_1^\alpha]G_1\mathrm{d}S$$

$$= \begin{cases} -\mathbf{E}_1^\alpha(r'), \; r' \in C\bar{T}, \\ 0, \; r' \in T \end{cases} \tag{4.36}$$

$$\frac{1}{i\omega\mu_1} rot' rot' \int_S [\mathbf{n} \times \mathbf{E}_1^\alpha] G_1 \mathrm{d}S + rot' \int_S [\mathbf{n} \times \mathbf{H}_1^\alpha] G_1 \mathrm{d}S$$

$$= \begin{cases} -\mathbf{H}_1^\alpha(r'), \; r' \in C\bar{T}, \\ 0, \; r' \in T \end{cases} \tag{4.37}$$

are used for a monochromatic field, it is possible to derive simpler equations of the inverse problem not containing the normal components of **E** and **H** under the integral sign:

$$\mathbf{E}_1^\alpha(r') = rot' rot' \int_S \left\{ [\mathbf{n} \times \mathbf{H}_1^\alpha] \left(\frac{G_2}{\sigma_2^*} - \frac{G_1}{\sigma_1^*} \right) + \frac{G_2}{\sigma_1^*} [\mathbf{n} \times \mathbf{H}^H] \right\} \mathrm{d}s$$

$$+ rot' \int_S \{ [\mathbf{n} \times \mathbf{E}_1^\alpha](G_2 - G_1) + G_2 [\mathbf{n} \times \mathbf{E}^H] \} \, \mathrm{d}s \tag{4.38}$$

$$\mathbf{H}_1^\alpha(r') = \frac{1}{i\omega} rot' rot' \int_S \left\{ [\mathbf{n} \times \mathbf{E}_1^\alpha] \left(\frac{G_2}{\mu_2} - \frac{G_1}{\mu_1} \right) + \frac{G_2}{\mu_2} [\mathbf{n} \times \mathbf{E}^H] \right\} \mathrm{d}s$$

$$+ rot' \int_S \{ [\mathbf{n} \times \mathbf{H}_1^\alpha](G_2 - G_1) + G_2 [\mathbf{n} \times \mathbf{H}^H] \} \, \mathrm{d}s. \tag{4.39}$$

Equations (4.32), (4.33), (4.38), (4.39) are TIP equations (relative to the boundary S) for a monochromatic field. In their numerical solution it is possible to use the algorithm formulated in Chapter 3.

The case $\sigma_2 = \infty$ (the domain T is a perfect conductor) is of a special interest. Then at the boundary S, we have

$$(\mathbf{H}_1, \mathbf{n}) = 0, \; [\mathbf{E}_1, \mathbf{n}] = 0. \tag{4.40}$$

Relations (4.40) are TIP functional equations for determining the surface of a perfect conductor, which, in turn, is a good zero approximation in solving (4.32), (4.33), (4.38), (4.39). We also note that in (Zhdanov, 1984) the theory of Stratton-Chu integrals for monochromatic fields is generalized for the case of inhomogeneous media (a magnetically homogeneous medium with a stipulated piecewise-continuous distribution of conductivity $\sigma^*(r)$ is considered, and the domain T is bounded by a smooth surface S). An expression was derived for the electromagnetic field on the basis of its values at the surface using the Green electromagnetic tensors:

$$\mathbf{E}(r') = \int\limits_S \left\{ i\omega\mu\hat{G}^e(r'|r)[\mathbf{n}\times\mathbf{H}] + [\tilde{\nabla}\times\widetilde{\hat{G}^e(r'|r)}][\mathbf{n}\times\mathbf{E}]\right\} \mathrm{d}s \qquad (4.41)$$

$$\mathbf{H}(r') = \int\limits_S \left\{ i\omega\mu\hat{G}^m(r'|r)[\mathbf{n}\times\mathbf{H}] + [\tilde{\nabla}\times\widetilde{\hat{G}^m(r'|r)}][\mathbf{n}\times\mathbf{E}]\right\} \mathrm{d}s \qquad (4.42)$$

where $\hat{G}^e, \hat{G}^m$ are tensor functions of electric and magnetic types, $r' \in T$. If $r' \in C\bar{T}$, the integrals in (4.31), (4.32) are equal to zero. In the case of an arbitrary medium, the determination of $\hat{G}^e$ and $\hat{G}^m$ is a rather complex problem. However, for a number of models (such as horizontally layered medium with an embedded body), this problem may be solved, and, therefore, it is possible to write the TIP equation.

For horizontal-layer medium there are analogs of Stratton-Chu formulas (Zakharov and Il'in, 1970).

Remark 1. If the domain T is bounded by a cylindrical surface S whose generatrix is parallel to the Y axis and the electromagnetic field $\mathbf{E}, \mathbf{H}$ is homogeneous along this axis, the problem is reduced to the two-dimensional one. The two-dimensional equations retain the structure of the three-dimensional equations; integration is carried out along a closed contour and

$$G(r'|r) = -\frac{i}{4}H_0^{(1)}(k^*|r-r'|)$$

where $H_0^{(1)}$ is the zero-order Hankel function of the first kind.

4.4. THE TIP EQUATIONS FOR QUASI-STATIONARY FIELD

In the case of a quasi-stationary field, the Stratton-Chu integrals have the form

$$\int_{-\infty}^{t'}\int_S \left\{(\mathbf{n},\mathbf{E}_2)\,\mathrm{grad}\,G_2^d + [\mathbf{n}\times\mathbf{E}_2]\times\mathrm{grad}\,G_2^d + \mu_2[\mathbf{n}\times\mathbf{H}_2]\frac{\partial G_2^d}{\partial t}\right\}\mathrm{d}S\mathrm{d}t$$
$$= \begin{cases}\mathbf{E}_2(r',t'),\ r'\in T\\ 0,\ r'\in C\bar{T}\end{cases} \tag{4.43}$$

$$\int_{-\infty}^{t'}\int_S \left\{(\mathbf{n},\mathbf{H}_2)\,\mathrm{grad}\,G_2^d + [\mathbf{n}\times\mathbf{H}_2]\times\mathrm{grad}\,G_2^d + \sigma_2[\mathbf{n}\times\mathbf{E}_2]\frac{\partial G_2^d}{\partial t}\right\}\mathrm{d}S\mathrm{d}t$$
$$= \begin{cases}\mathbf{H}_2(r',t'),\ r'\in T\\ 0,\ r'\in C\bar{T}\end{cases} \tag{4.44}$$

$$\int_{-\infty}^{t'}\int_S \left\{(\mathbf{n},\mathbf{E}_1^\alpha)\,\mathrm{grad}\,G_1^d + [\mathbf{n}\times\mathbf{E}_1^\alpha]\times\mathrm{grad}\,G_1^d + \mu_1[\mathbf{n}\times\mathbf{H}_1^\alpha]\frac{\partial G_1^d}{\partial t}\right\}\mathrm{d}S\mathrm{d}t$$
$$= \begin{cases}\mathbf{E}_1^\alpha(r',t'),\ r'\in C\bar{T}\\ 0,\ r'\in T\end{cases} \tag{4.45}$$

$$\int_{-\infty}^{t'}\int_S \left\{(\mathbf{n},\mathbf{H}_1^\alpha)\,\mathrm{grad}\,G_1^d + [\mathbf{n}\times\mathbf{H}_1^\alpha]\times\mathrm{grad}\,G_1^d + \sigma_1[\mathbf{n}\times\mathbf{E}_1^\alpha]\frac{\partial G_1^d}{\partial t}\right\}\mathrm{d}S\mathrm{d}t$$
$$= \begin{cases}\mathbf{H}_1^\alpha(r',t'),\ r'\in C\bar{T}\\ 0,\ r'\in T\end{cases} \tag{4.46}$$

$$G^d(r,t\mid r',t') = \frac{-\sqrt{\mu\sigma}}{8\pi^{3/2}(t'-t)^{3/2}}\exp\left[-\frac{\mu\sigma\mid r-r'\mid^2}{4(t'-t)}\right]æ(t'-t). \tag{4.47}$$

Here G^d is the Green function for the diffusion equation, and

$$æ(t) = \begin{cases}0,\ t\le 0,\\ 1,\ t>0\end{cases} \tag{4.48}$$

is the Heaviside function.

With (4.43)-(4.46), it is possible to derive the TIP equation for a quasi-stationary field:

$$\mathbf{E}_1^a(r',t')=\int_{-\infty}^{t'}\int_S\Big\{[\mathbf{n}\times\mathbf{E}_1^a]\times grad(G_2^d-G_1^d)+[\mathbf{n}\times\mathbf{H}_1^a](\mu_2\frac{\partial G_2^d}{\partial t}-\mu_1\frac{\partial G_1}{\partial t})$$

$$+[\mathbf{n}\times\mathbf{E}^H]\times\operatorname{grad}G_2^d+\mu_2[\mathbf{n}\times\mathbf{H}^H]\frac{\partial G_2^d}{\partial t} \tag{4.49}$$

$$+(\mathbf{n},\mathbf{E}_1^a)\left[\operatorname{grad}(\frac{\sigma_1}{\sigma_2}G_2^d-G_1^d)\right]+\frac{\sigma_1\operatorname{grad}G_2^d}{\sigma_2}(\mathbf{n},\mathbf{E}^H)]\Big\}\,dsdt$$

$$\mathbf{H}_1^a(r',t')=\int_{-\infty}^{t'}\int_S\Big\{(\mathbf{n},\mathbf{H}_1^a)\left[\operatorname{grad}(\frac{\mu_1}{\mu_2}G_2^d-G_1^d)\right]+\frac{\mu_1}{\mu_2}(\mathbf{H}^H,\mathbf{n})\operatorname{grad}G_2^d$$

$$+[\mathbf{n}\times\mathbf{H}_1^a]\times grad(G_2^d-G_1^d)+[\mathbf{n}\times\mathbf{E}_1^a](\sigma_2G_2^d-\sigma_1G_1^d) \tag{4.50}$$

$$+[\mathbf{n}\times\mathbf{H}^H]\times\operatorname{grad}G_2^d+[\mathbf{n}\times\mathbf{E}^H]\sigma_2G^d\Big\}\,\mathrm{d}S\mathrm{d}t.$$

For $\mathbf{E}$ it is possible to derive a simpler equation, taking rot' from both sides of (4.49):

$$\mathbf{E}_1^a(r',t')=\int_{-\infty}^{t'}\Big\{rot'rot'\int_S\left([\mathbf{n}\times\mathbf{H}_1^\alpha]\left(\frac{G_2^d}{\sigma_2}-\frac{G_1^d}{\sigma_1}\right)+\frac{G_2^d}{\sigma_1}[\mathbf{n}\times\mathbf{H}^H]\right)\mathrm{d}S$$

$$+rot'\int_S\left([\mathbf{n}\times\mathbf{E}_1^\alpha]\left(G_2^d-G_1^d\right)+G_2^d[\mathbf{n}\times\mathbf{E}^H]\right)\mathrm{d}S\Big\}\,\mathrm{d}t. \tag{4.51}$$

4.5. THE TIP EQUATIONS FOR THE CASE $\sigma_2 = 0$

In the case of the wave field, the following expressions for **E** and **H** hold (Zhdanov, 1984):

$$-\frac{1}{4\pi\sqrt{\mu_2\varepsilon_2}}\left\{-\operatorname{grad}'\int_S \frac{\mathbf{n}\cdot\widetilde{\mathbf{E}}_2}{|r-r'|}\mathrm{d}S + rot'\int_S \frac{\mathbf{n}\times\widetilde{\mathbf{E}}_2}{|r-r'|}\mathrm{d}S \right.$$

$$\left. -\mu_2\frac{\partial}{\partial t'}\int_S \frac{\mathbf{n}\times\widetilde{\mathbf{H}}_2}{|r-r'|}\mathrm{d}S\right\} = \begin{cases}\mathbf{E}_2(r',t'),\ r'\in T\\ 0,\ r'\in C\bar{T}\end{cases} \tag{4.52}$$

$$-\frac{1}{4\pi\sqrt{\mu_2\varepsilon_2}}\left\{-\operatorname{grad}'\int_S \frac{\mathbf{n}\cdot\widetilde{\mathbf{H}}_2}{|r-r'|}\mathrm{d}S + rot'\int_S \frac{\mathbf{n}\times\widetilde{\mathbf{H}}_2}{|r-r'|}\mathrm{d}S \right.$$

$$\left. +\varepsilon_2\frac{\partial}{\partial t'}\int_S \frac{\mathbf{n}\times\widetilde{\mathbf{E}}_2}{|r-r'|}\mathrm{d}S\right\} = \begin{cases}\mathbf{H}_2(r',t'),\ r'\in T\\ 0,\ r'\in C\bar{T}\end{cases} \tag{4.53}$$

where $\widetilde{\mathbf{E}}_2 = \mathbf{E}_2(r, t' - \sqrt{\mu_2\varepsilon_2}|r-r'|)$ and $\widetilde{\mathbf{H}}_2 = \mathbf{H}_2(r, t' - \sqrt{\mu_2\varepsilon_2}|r-r'|)$ are the quantities of electric and magnetic fields in observed moments of time. Substituting $\mathbf{E}_1$ and $\mathbf{H}_1$ into (4.52) and (4.53) instead of $\mathbf{E}_2$ and $\mathbf{H}_2$ (at $r' \in C\overline{T}$), we find the equations of the theoretical inverse problem (relative to the boundary S) for the wave electromagnetic field:

$$-\frac{1}{4\pi\sqrt{\mu_2\varepsilon_2}}\left\{-\operatorname{grad}'\int_S \frac{\mathbf{n}\cdot\widetilde{\mathbf{E}}_1\varepsilon_1}{|r-r'|\varepsilon_2}dS + rot'\int_S \frac{\mathbf{n}\times\widetilde{\mathbf{E}}_1}{|r-r'|}\mathrm{d}S \right.$$

$$\left. -\mu_2\frac{\partial}{\partial t'}\int_S \frac{\mathbf{n}\times\widetilde{\mathbf{H}}_1}{|r-r'|}\mathrm{d}S\right\} = 0 \tag{4.54}$$

$$-\frac{1}{4\pi\sqrt{\mu_2\varepsilon_2}}\left\{-\operatorname{grad}'\int_S \frac{\mathbf{n}\cdot\widetilde{\mathbf{H}}_1\mu_1}{|r-r'|\mu_2}\mathrm{d}S + rot'\int_S \frac{\mathbf{n}\times\widetilde{\mathbf{H}}_1}{|r-r'|}\mathrm{d}S \right.$$

$$\left. +\varepsilon_2\frac{\partial}{\partial t'}\int_S \frac{\mathbf{n}\times\widetilde{\mathbf{E}}_1}{|r-r'|}\mathrm{d}S\right\} = 0. \tag{4.55}$$

For the quasi-stationary field $\mathbf{E}_2, \mathbf{H}_2$ and $\sigma_2 = 0$, we have

$$\left.\begin{array}{r}\mathbf{E}_2(r'),\ r' \in T \\ 0,\quad r' \in C\bar{T}\end{array}\right\} = \frac{1}{4\pi}\int\limits_S \left\{(\mathbf{n}, \mathbf{E}_2)\,\mathrm{grad}\,\frac{1}{|r-r'|} + [\mathbf{n}\times\mathbf{E}_2]\times\mathrm{grad}\,\frac{1}{|r-r'|}\right\}\mathrm{d}S \tag{4.56}$$

$$\left.\begin{array}{r}\mathbf{H}_2(r'),\ r' \in T \\ 0,\quad r' \in C\bar{T}\end{array}\right\} = \frac{1}{4\pi}\int\limits_S \left\{(\mathbf{n}, \mathbf{H}_2)\,\mathrm{grad}\,\frac{1}{|r-r'|} + [\mathbf{n}\times\mathbf{H}_2]\times\mathrm{grad}\,\frac{1}{|r-r'|}\right\}\mathrm{d}S. \tag{4.57}$$

Substituting boundary conditions (4.18) to (4.56) and (4.57) when $r' \in C\bar{T}$, we have corelations for seeking boundary insulator:

$$\frac{1}{4\pi}\int\limits_S \left\{\frac{\varepsilon_1}{\varepsilon_2}(\mathbf{n}, \mathbf{E}_1)\,\mathrm{grad}\,\frac{1}{|r-r'|} + [\mathbf{n}\times\mathbf{E}_1]\times\mathrm{grad}\,\frac{1}{|r-r'|}\right\}\mathrm{d}S = 0 \tag{4.58}$$

$$\frac{1}{4\pi}\int\limits_S \left\{\frac{\mu_1}{\mu_2}(\mathbf{n}, \mathbf{H}_1)\,\mathrm{grad}\,\frac{1}{|r-r'|} + [\mathbf{n}\times\mathbf{H}_1]\times\mathrm{grad}\,\frac{1}{|r-r'|}\right\}\mathrm{d}S = 0. \tag{4.59}$$

Remarks 2. In the case of direct-current prospecting, the theoretical inverse problem for the domain T with $\sigma_2 = 0$ is reduced to determining the inclusion boundary S from the relation

$$\left.\frac{\partial U}{\partial n}\right|_S = 0, \quad \text{i.e.} \quad (\mathrm{grad}\,\mathbf{U}, n) = 0,$$

where U is the electric (total) potential, and $\mathbf{n}$ is the normal vector to the surface S.

If one needs to determine an isolating interface described by an equation of the form $z = z(x, y)$, then relation (17) may be expressed as follows:

$$\frac{\partial U}{\partial x}\frac{\partial z}{\partial x} + \frac{\partial U}{\partial y}\frac{\partial z}{\partial y} - \frac{\partial U}{\partial z} = 0.$$

4.6. THE TIP EQUATIONS FOR ARBITRARY ELECTROMAGNETIC FIELD

If the functions **E** and **H** satisfy telegrapher's equation, then the Stratton-Chu formulas have the form

$$\int_S \int_{-\infty}^{t'-\sqrt{\mu_2\varepsilon_2}|r-r'|} \{(\mathbf{n},\mathbf{E}_2)\,\mathrm{grad}\,G_2^t + [\mathbf{n}\times\mathbf{E}_2]\times\mathrm{grad}\,G_2^t + \mu_2[\mathbf{n}\times\mathbf{H}_2]\frac{\partial G_2^t}{\partial t}\}\,\mathrm{d}t\mathrm{d}S = \begin{cases}\mathbf{E}_2(r',t'),\ r'\in T,\\ 0,\ r'\in C\bar{T}\end{cases} \tag{4.60}$$

$$\int_S \int_{-\infty}^{t'-\sqrt{\mu_2\varepsilon_2}|r-r'|} \{(\mathbf{n},\mathbf{H}_2)\,\mathrm{grad}\,G_2^t + [\mathbf{n}\times\mathbf{H}_2]\times\mathrm{grad}\,G_2^t + \sigma_2[\mathbf{n}\times\mathbf{E}_2]G_2^t - \varepsilon_2[\mathbf{n}\times\mathbf{E}_2]\frac{\partial G_2^t}{\partial t}\}\,\mathrm{d}t\mathrm{d}S = \begin{cases}\mathbf{H}_2(r',t'),\ r'\in T,\\ 0,\ r'\in C\bar{T}\end{cases} \tag{4.61}$$

where G_2 is the fundamental solution for telegrapher's equation

$$G^t(r,t\mid r',t') = \frac{\exp[-\sigma(t'-t)/2\varepsilon]}{4\pi\sqrt{\mu\varepsilon}\mid r-r'\mid}\left\{\delta\left(\frac{t'-t-\sqrt{\mu\varepsilon}\mid r-r'\mid}{\sqrt{\mu\varepsilon}}\right)\right.$$

$$+\frac{1}{2}\sqrt{\frac{\mu\sigma^2}{\varepsilon}}\frac{\mid r-r'\mid}{\sqrt{\mid r-r'\mid^2-(t'-t)^2/\mu\varepsilon}}J_1\left[\frac{1}{2}\sqrt{\frac{\mu\sigma^2}{\varepsilon}}\sqrt{\mid r-r'\mid^2+(t'-t)^2/\mu\varepsilon}\right] \tag{4.62}$$

$$\left.\times æ\left(\frac{t'-t-\sqrt{\mu\varepsilon}\mid r-r'\mid}{\sqrt{\mu\varepsilon}}\right)\right\},$$

J_1 is the Bessel function, and æ is the Heaviside function.

Taking into account the similar representations for $\mathbf{E}_1$ and $\mathbf{H}_1$ when, $r'\in C\bar{T}$, we obtain the TIP equation for the nonstationary electromagnetic field (in the following case):

$$\mathbf{E}_1^a(r',t') = \int_S \int_{-\infty}^{t'-\sqrt{\mu\varepsilon}|r-r'|} \left\{(\mathbf{n},\mathbf{E}_1^a)\,\mathrm{grad}\left(\frac{\varepsilon_1}{\varepsilon_2}G_2^t - G_1^t\right)\right.$$

$$+\frac{\operatorname{grad} G_2^t}{\varepsilon_2}[\eta+\varepsilon_1(\mathbf{E}_1^H,\mathbf{n})]+[\mathbf{n}\times\mathbf{E}_1^a]\times\operatorname{grad}(G_2^t-G_1^t)+[\mathbf{n}\times\mathbf{H}^H]\mu_2\frac{\partial G_2^t}{\partial t} \tag{4.63}$$

$$+[\mathbf{n}\times\mathbf{E}^H]\times\operatorname{grad} G_2^t+[\mathbf{n}\times\mathbf{H}_1^a]\left(\mu_2\frac{\partial G_2^t}{\partial t}-\mu_1\frac{\partial G_1^t}{\partial t}\right)\Bigg\}\,dtdS,$$

$$\mathbf{H}_1^a(r',t')=\int\limits_S\int\limits_{-\infty}^{t'-\sqrt{\mu\varepsilon}|r-r'|}\Bigg\{(\mathbf{n},\mathbf{H}_1^a)\operatorname{grad}\left(\frac{\mu_1}{\mu_2}G_2^t-G_1^t\right)$$

$$+[\mathbf{n}\times\mathbf{H}_1^a]grad\left(G_2^t-G_1^t\right)+[\mathbf{n}\times\mathbf{H}^H]\operatorname{grad} G_2^t+[\mathbf{n}\times\mathbf{E}^H]\left(\sigma_2^*G_2^t-\varepsilon_2\frac{\partial G_2^t}{\partial t}\right)$$

$$+\frac{\mu_1}{\mu_2}(\mathbf{H}_1^H,\mathbf{n})\operatorname{grad} G_2^t+[\mathbf{n}\times\mathbf{E}_1^a](\sigma_2^*G_2^t-\sigma_1^*G_1^t)-$$

$$-[\mathbf{n}\times\mathbf{E}_1^a]\left(\varepsilon_2\frac{\partial G_2^t}{\partial t}-\varepsilon_1\frac{\partial G_1^t}{\partial t}\right)\Bigg\}\,dtdS. \tag{4.64}$$

If there are currents and charges in the domain T then relations (4.63) and (4.64) become as follows

$$\mathbf{E}_1^a(r',t')=\int\limits_S\int\limits_{-\infty}^{t'-\sqrt{\mu\varepsilon}|r-r'|}\Bigg\{(\mathbf{n},\mathbf{E}_1^a)\operatorname{grad}\left(\frac{\varepsilon_1}{\varepsilon_2}G_2^t-G_1^t\right)$$

$$+\frac{\operatorname{grad} G_2^t}{\varepsilon_2}[\eta+\varepsilon_1(\mathbf{E}^H,\mathbf{n})]+[\mathbf{n}\times\mathbf{E}^H]\times\operatorname{grad} G_2^t+[\mathbf{n}\times\mathbf{E}_1^a]\times\operatorname{grad}(G_2^t-G_1^t)$$

$$+[\mathbf{n}\times\mathbf{H}^H]\mu_2\frac{\partial G_2^t}{\partial t}+[\mathbf{n}\times\mathbf{H}_1^a]\left(\mu_2\frac{\partial G_2^t}{\partial t}-\mu_1\frac{\partial G_1^t}{\partial t}\right)\Bigg\}\,dSdt$$

$$-\iiint\limits_T\int\limits_{-\infty}^{t'-\sqrt{\mu\varepsilon}|r-r'|}\left\{\frac{\widetilde{q^e}}{\varepsilon}\operatorname{grad} G^t+\mu\mathbf{j}^e\frac{\partial G^t}{\partial t}-\mathbf{j}^m\times\operatorname{grad} G^t\right\}dtdV \tag{4.65}$$

$$\mathbf{H}_1^a(r',t') = \int\limits_S \int\limits_{-\infty}^{t'-\sqrt{\mu\varepsilon}|r-r'|} \left\{ (\mathbf{n}, \mathbf{H}_1^a) \operatorname{grad} \left(\frac{\mu_1}{\mu_2} G_2^t - G_1^t \right) \right.$$

$$+ \frac{\mu_1}{\mu_2} (\mathbf{H}^H, \mathbf{n}) \operatorname{grad} G_2^t + [\mathbf{n} \times \mathbf{H}_1^a] \times \operatorname{grad} (G_2^t - G_1^t)$$

$$+ [\mathbf{n} \times \mathbf{H}^H] \operatorname{grad} G_2^t + [\mathbf{n} \times \mathbf{E}^H] \left(\sigma_2 G_2^t - \varepsilon_2 \frac{\partial G_2^t}{\partial t} \right)$$

$$\left. + [\mathbf{n} \times \mathbf{E}_1^a](\sigma_2 G_2^t - \sigma_1 G_1^t) - [\mathbf{n} \times \mathbf{E}_1^a] \left(\varepsilon_2 \frac{\partial G_2^t}{\partial t} - \varepsilon_1 \frac{\partial G_1^t}{\partial t} \right) \right\} \mathrm{d}t\mathrm{d}s$$

$$- \iiint\limits_T \int\limits_{-\infty}^{t'-\sqrt{\mu\varepsilon}|r-r'|} \left\{ \frac{\widetilde{q^m}}{\varepsilon} \operatorname{grad} G^t - \sigma \mathbf{j}^m \frac{\partial G^t}{\partial t} \right.$$

$$\left. + \varepsilon \mathbf{j}^m \frac{\partial G^t}{\partial t} - \mathbf{j}^e \times \operatorname{grad} G^t \right\} \mathrm{d}t\mathrm{d}V. \tag{4.66}$$

Remark 3. In principle, for a TIP solution, it is possible to use integral representations for the functions $\mathbf{E}_2$ and $\mathbf{H}_2$ when $r' \in C\bar{T}$, replacing in the integral $\mathbf{E}_2$ and $\mathbf{H}_2$ by $\mathbf{E}_1^a$ and $\mathbf{H}_1^a$ from conditions (4.18). Then the TIP is reduced to a variational problem. For example, instead of (4.63), we have

$$\int\limits_S \int\limits_{-\infty}^{t'-\sqrt{\mu_2\varepsilon_2}|r-r'|} \left\{ \frac{\mu_1}{\mu_2} (\mathbf{H}_1, \mathbf{n}) \operatorname{grad} G_2^t + [\mathbf{n} \times \mathbf{H}_1] \times \operatorname{grad} G_2^t \right.$$

$$\left. + \sigma_2 [\mathbf{n} \times \mathbf{E}_1] G_2^t - \varepsilon_2 [\mathbf{n} \times \mathbf{E}_1] \frac{\partial G_2^t}{\partial t} \right\} \mathrm{d}t\mathrm{d}S = 0. \tag{4.67}$$

Similarly, instead of (4.33), we obtain

$$\int\limits_S \left\{ (\mathbf{n}, \mathbf{H}_1) \frac{\mu_1}{\mu_2} \operatorname{grad} G_2 + [\mathbf{n} \times \mathbf{H}_1] \times \operatorname{grad} G_2 + \sigma_2^* [\mathbf{n} \times \mathbf{E}_1] G_2 \right\} \mathrm{d}s = 0 \tag{4.68}$$

etc.

However, for variational problems there is no uniqueness of solution for fixed $\sigma_2, \varepsilon_2, \mu_2$ (that is, the surface S is not determined uniquely).

4.7. ALGORITHM FOR SOLVING THE TIP EQUATIONS IN THE CLASS OF STELLAR BODIES

Bodies stellar in relation to a certain internal point constitute a broad class important in practical terms. We shall construct a spherical system of coordinates with the center at that point.

Making in (4.32′) a spherical substitution of the variables, we obtain

$$\mathbf{E}_1^{\alpha}(r') = \iint_0^{2\pi} \iint_0^{\pi} U_{\sigma_2,\varepsilon_2}(\rho(\Theta,\phi),\Theta,\phi,r')\,\mathrm{d}\Theta\,\mathrm{d}\phi \tag{4.69}$$

where U is function under the integral in (4.32′), $\rho(\Theta,\phi)$ is the right-hand side of the equation of the body surface S $-\rho = \rho(\Theta,\phi)$ in spherical coordinates. Representation (4.69) may be considered as the explicit equation relative to the function $\rho(\Theta,\phi)$.

The right-hand side of the equation of the surface of the body T may be represented by a double Fourier series:

$$\rho(\Theta,\phi) = \sum_{k=-\infty}^{\infty}\sum_{j=-\infty}^{\infty} a_{kj} e^{i(k\Theta + j\phi)}. \tag{4.70}$$

We shall therefore seek the solution of (4.69) in the form of a segment of the double Fourier series

$$\rho(\Theta,\phi) = \sum_{k=-n}^{n}\sum_{j=-m}^{m} a_{kj} e^{i(k\Theta + j\phi)} \tag{4.71}$$

minimizing the functional

$$f(\gamma) = \sum_{i\in M} \left[\mathbf{E}_1^{\alpha}(r'_i) - \mathbf{E}_1^{\alpha,nm}(r'_i;\gamma)\right]^2 \tag{4.72}$$

where the points r'_i belong to the information carrier which in the case of TIP can be chosen; γ is the vector of the coefficients of the function $\rho(\Theta,\phi)$, for which minimization is performed; and $\mathbf{E}_1^{\alpha,nm}$ is the right-hand side of (4.69) substituting ρ by ρ_{nm}.

It is known that the function $\rho(\Theta,\phi)$, which defines the boundary of the one-connected three-dimensional domain, must have the following property:

$$\rho(0,\phi) = \text{const}, \qquad \rho(\pi,\phi) = \text{const}. \tag{4.73}$$

From the condition above, the relation for the coefficients ρ_{nm} it follows that:

$$\sum_k a_{kj} = 0, \qquad \sum_{k_1} a_{k_1 j} = 0, j \in [-m, m] \tag{4.74}$$

$$k, k_1 \in [-n, n], \qquad k = \pm 1, \pm 3, \cdots, \\ k_1 = \pm 2, \pm 4, \cdots. \tag{4.75}$$

We must find thus the minimum of the functional f and subject to equality-type constraints. Practically, we add to the functional f the functional f_1

$$f_1 = \sum_j P_j \left(\left| \sum_k a_{kj} \right|^2 + \left| \sum_{k_1} a_{k_1 j} \right|^2 \right)$$

where P_j are the penalty coefficients. Regularization of (4.69) is done in (Blokh, 1975) by using an analog of smoothing first-order Tikhonov normalizer (Gakhov, 1963)

$$\Omega(\rho) = \iint\limits_0^{2\pi} \iint\limits_0^{\pi} (|\rho'_\Theta|^2 + |\rho\prime_\phi|^2) d\Theta d\phi = 2\pi^2 \sum_{k=-n}^{n} \sum_{j=-m}^{m} (k^2 + j^2)|a_{kj}|^2. \tag{4.76}$$

In solving (4.69), the functional

$$f + f_1 + \alpha\Omega \tag{4.77}$$

was minimized, where the regularization parameter α was determined from the discrepancy. The Powell method was used for minimization (Glasko *et al.*, 1970); the integral in (4.69) was computed by using the formulas of a high trigonometric accuracy, which in this case are the formulas with equal weights and equidistant nodes. A sphere of a minimum radius including all the singularities $\mathbf{E}_1^\alpha$ was taken as an initial approximation. The solution was then defined for (4.69) (which includes the conductivity σ_2 as parameter, making it possible to construct an whole family of equivalent bodies generating with different conductivities the similar field) for a certain value of $\sigma_2 = \sigma_0$. At each subsequent step the body T_{n-1} was taken as the initial approximation for the conductivity $\sigma^n < \sigma^{n-1}$. A sphere known *a priori* to contain the solution was chosen for the information carrier.

Numerical examples

There are some theoretical examples ($\mu_1 = \mu_2 = \mu_0$, $\mu_0 = 4\pi \cdot 10^{-7}$ H/m.)

Example 1. Figure 4.1 shows the cross-sections cut off by the coordinate planes on TIP solutions of the Helmholts equation for the function $\mathbf{E}_1^\alpha = (Q_{1x}\frac{e^{ik_2r_1}}{r_1} + Q_{2x}\frac{e^{ik_2r_2}}{r_2} + Q_{3x}\frac{e^{ik_2r_3}}{r_3},\ Q_{1y}\frac{e^{ik_2r_1}}{r_1} + Q_{2y}\frac{e^{ik_2r_2}}{r_2} + Q_{3y}\frac{e^{ik_2r_3}}{r_3},$ $Q_{1z}\frac{e^{ik_2r_1}}{r_1} + Q_{2z}\frac{e^{ik_2r_2}}{r_2} + Q_{3z}\frac{e^{ik_2r_3}}{r_3})$, $\mathbf{E}^S(P) = (Q_{4x}\frac{e^{ik_1r_4}}{r_4},\ Q_{4y}\frac{e^{ik_1r_4}}{r_4},\ Q_{4z}\frac{e^{ik_1r_4}}{r_4})$, P_1, P_2,
$P_3 \in T^+$, $P_4, P \in T^-$, $r_i = |PP_i|$, $i = \overline{1,4}$, $P_4(0,0,9)$, $Q_{1x} = Q_{1y} = Q_{1z} = 1$, $Q_{2x} = Q_{2y} = Q_{2z} = 1$, $Q_{3x} = Q_{3y} = Q_{3z} = -2$, $Q_{4x} = Q_{4y} = Q_{4z} = -5$, $\omega = 4\pi \cdot 10000$.

Example 2. Figure 2 shows the cross-sections cut off by the coordinate planes on the TIP solutions of the Helmholtz equation for function $\mathbf{E}_1^\alpha = (Q_{1x}\frac{e^{ik_2r_1}}{r_1} + Q_{2x}\frac{e^{ik_2r_2}}{r_2} + Q_{3x}\frac{e^{ik_2r_3}}{r_3},\ Q_{1y}\frac{e^{ik_2r_1}}{r_1} + Q_{2y}\frac{e^{ik_2r_2}}{r_2} + Q_{3y}\frac{e^{ik_2r_3}}{r_3},$ $Q_{1z}\frac{e^{ik_2r_1}}{r_1} + Q_{2z}\frac{e^{ik_2r_2}}{r_2} + Q_{3z}\frac{e^{ik_2r_3}}{r_3})$, $\mathbf{E}^S(P) = (Q_{4x}\frac{e^{ik_1r_4}}{r_4},\ Q_{4y}\frac{e^{ik_1r_4}}{r_4},\ Q_{4z}\frac{e^{ik_1r_4}}{r_4})$, P_1, P_2,
$P_3 \in T^+$, $P_4, P \in T^-$, $r_i = |PP_i|$, $i = \overline{1,4}$, $P_4(0,0,9)$, $Q_{1x} = Q_{1y} = Q_{1z} = 1$, $Q_{2x} = Q_{2y} = Q_{2z} = 2$, $Q_{3x} = Q_{3y} = Q_{3z} = -3$, $Q_{4x} = Q_{4y} = Q_{4z} = -5$, $\omega = 4\pi \cdot 10000$.

Example 3. Figure 3 shows the cross-sections cut off by the coordinate planes on the TIP solutions of the Helmholtz equation for function $\mathbf{E}_1^\alpha = (Q_{1x}\frac{e^{ik_2r_1}}{r_1} + Q_{2x}\frac{e^{ik_2r_2}}{r_2} + Q_{3x}\frac{e^{ik_2r_3}}{r_3},\ Q_{1y}\frac{e^{ik_2r_1}}{r_1} + Q_{2y}\frac{e^{ik_2r_2}}{r_2} + Q_{3y}\frac{e^{ik_2r_3}}{r_3},$ $Q_{1z}\frac{e^{ik_2r_1}}{r_1} + Q_{2z}\frac{e^{ik_2r_2}}{r_2} + Q_{3z}\frac{e^{ik_2r_3}}{r_3})$, $\mathbf{E}^S(P) = (Q_{4x}\frac{e^{ik_1r_4}}{r_4},\ Q_{4y}\frac{e^{ik_1r_4}}{r_4},\ Q_{4z}\frac{e^{ik_1r_4}}{r_4})$, P_1, P_2,
$P_3 \in T^+$, $P_4, P \in T^-$, $r_i = |PP_i|$, $i = \overline{1,4}$, $P_1(1,0,0)$, $P_2(-1.5,0,0)$, $P_3(0,0,1)$, $P_4(0,0,9)$, $Q_{1x} = Q_{1y} = Q_{1z} = 1$, $Q_{2x} = Q_{2y} = Q_{2z} = 2$, $Q_{3x} = Q_{3y} = Q_{3z} = -3$, $Q_{4x} = Q_{4y} = Q_{4z} = -5$, $\omega = 4\pi \cdot 10000$.

For all these examples the solutions of TIP for $\sigma_1/\sigma_2 = 1/5$, $\sigma_1/\sigma_2 = 1/10$ have been constructed.

Summary

Explicit equations of the theoretical inverse problems for electromagnetic fields were derived. We have considered telegrapher's, diffusion, and Helmholtz equations. The algorithms were developed and have been used for solving these equations in the case of monochromatic field. So, we have a technique for construction of the equivalent solutions of TIP and for studying principal questions: existence and uniqueness of solutions.

It is possible to develop methods for interpretation of variable electromagnetic fields which take into account the availability of equivalent solutions and the methods for approximation experimental data with the fields of singular sources.

4.8. CONCLUSIONS

We derived the explicit equations of the theoretical inverse problems for finding the surface of a three-dimensional insulator.

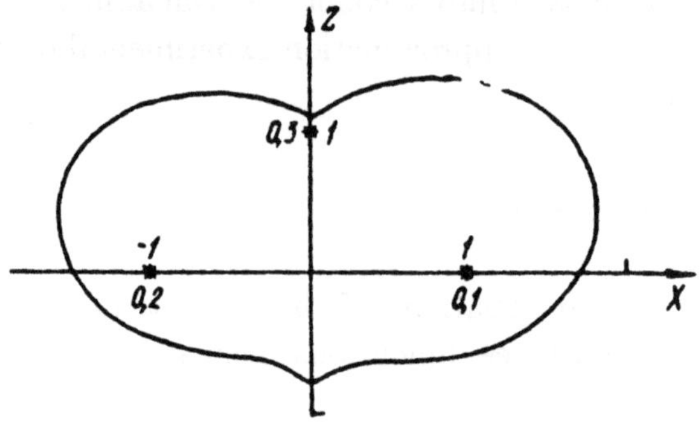

Figure 4.1

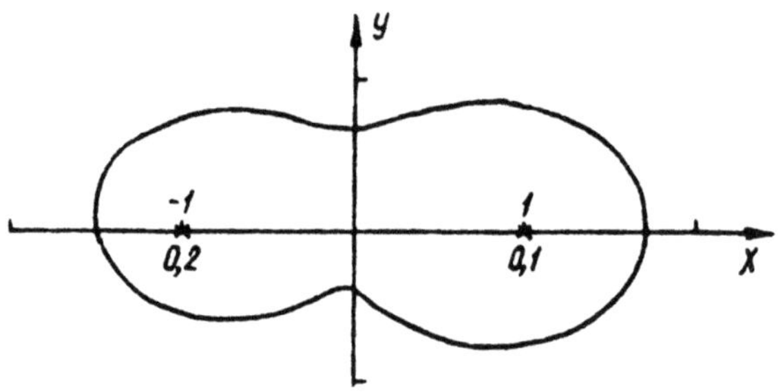

Figure 4.2

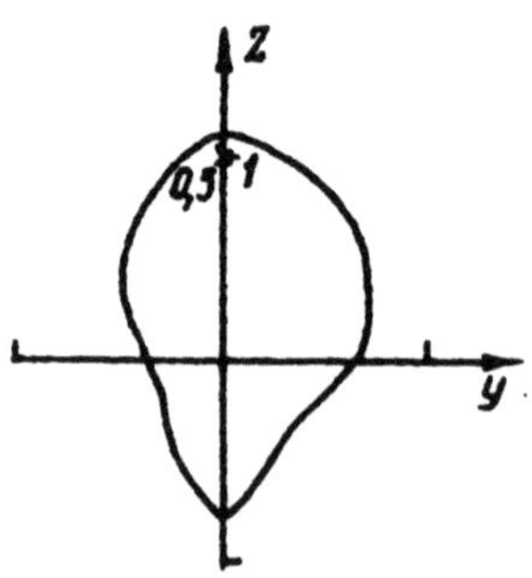

Figure 4.3

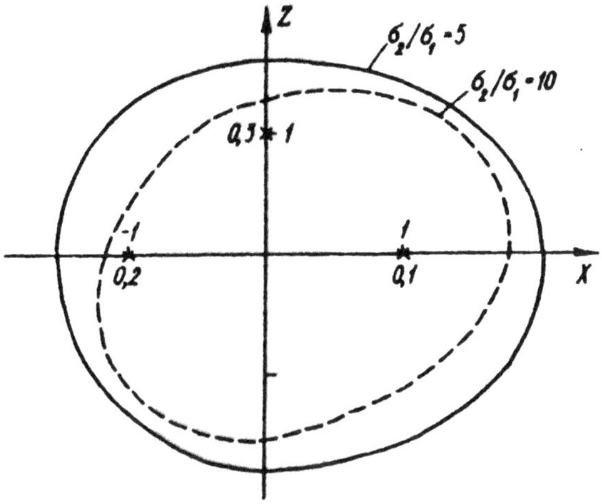

Figure 4.4

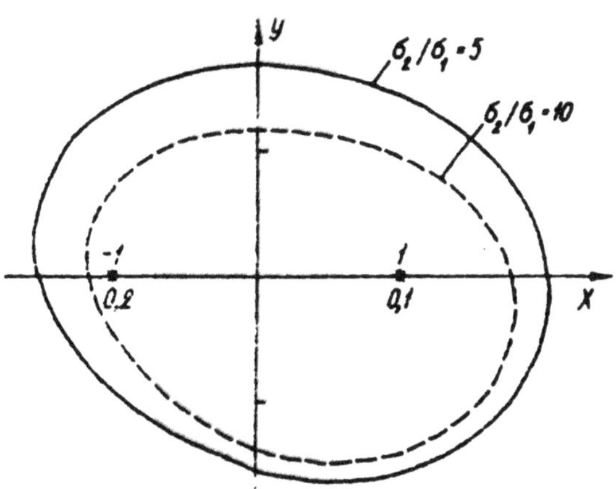

Figure 4.5

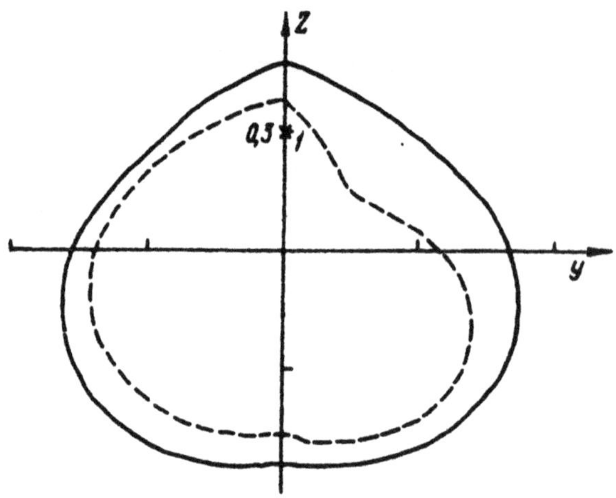

Figure 4.6

Bibliography

Atakhodzhaev, M. A. (1966). On uniqueness of solution of inverse mixed problem of the potential theory. *Sibirsk. Math. Journ.* **2**, 455–458 (in Russian).

Berdichevskii, M. N. and Zhdanov, M. S. (1975). Analysis of anomalies of a variable electromagnetic field in a surface of multi-layered horizontal inhomogeneous Earth. *Geomagnetic and Aeronomy* **15**(2), 372–374 (in Russian).

Berdichevskii, M. N. and Zhdanov, M. S. (1981). On extension of variable electromagnetic field in horizontal and homogeneous layers. *Geomagnethic Researches* **15**, 60–68 (in Russian).

Berdichevskii, M. N. and Zhdanov, M. S. (1981). *Interpretation of Anomalies of Variable Electromagnetic Field of the Earth.* Nedra, Moscow (in Russian).

Berdichevskii, M. N., Zhdanov, M. S., and Zhdanova, O. N. (1974). On possibility of division of anomalies of variable geomagnetic field in surface anomalies and depth anomalies. *Geomagnetics and Aeronomy* **14**(2), 136–143 (in Russian).

Blokh, Yu. I. (1975). On solution of the direct magnetic prospecting problem for the bodies of an arbitrary form taking account of demagnetization. *Izv. Vuzov. Geology and Prospecting* **1**, 100–108 (in Russian).

Blokh, Yu. I. (1981). Combined interpretation of magnetic prospecting data and of the nongrounded loop method when determining magnetic properties of magnetic stuff. *Izv. Akad. Nauk SSSR. Fizika Zemli* **11**, 74–84 (in Russian).

Blokh, Yu. I. (1987). Solution of the direct problem of magnetic prospecting for three-dimensional anisotropy geology objects taking account of demagnetization. *Izv. Akad. Nauk SSSR. Fizika Zemli* (in Russian).

Blokh, Yu. I., Garanskii, E. M., Dobrokhotova, I. A. *et al.*(1986). *Low Frequency Inductive Electroprospecting when Research and Prospecting Magnetic Stuff.* Nedra, Moscow (in Russian).

Cherednichenko, V. G. (1982). One problem of conjugation of analytic functions and the problem inverse to it. *Differential Equations* **18**(4), 682–689 (in Russian).

Cherednichenko, V. G. (1983). Extension with respect to a parameter of solution of a two-dimensional inverse potential problem. *Dokl. Akad. Nauk SSSR* **268**(2), 299–302 (in Russian).

Chudinova, A. A. (1965). Integral equation of the inverse electroprospecting problem. *Izv. Vuzov. Ser. Math.* **6**, 1080–1088 (in Russian).

Dmitriev, V. I. (1969). *Electromagnetic Fields in Inhomogeneous Media.* Izd. Mosc. Gos. Univ., Moscow (in Russian).

Dmitriev, V. I. and Zakharov, E. V. (1970). Method of solution of electrodynamic problems for inhomogeneous media. *Journ. Calc. Math. and Mech.* **10**(6), 1458–1464 (in Russian).

Dmitriev, V. I. and Zhdanov, M. S. (1982). Methods of solution of inverse geophysic problem. *Calc. Math. and Tech. in Prospecting Geophys* 89–105, Moscow (in Russian).

Eaton, P. A. (1989). 3-D electromagnetic inversion using integral equations. *Geophys. Prosp.* **37**, 407–426.

Filatov, V. A. (1968). A problem of numerical solution of inverse problem in the method of arfifitial magnetization. *Geology and Geophysics* **6**, 70–75 (in Russian).

Filatov, V. A. (1969a). On the inverse magnetic prospecting problem. *Dokl. Akad. Nauk SSSR* **186**(2), 1315–1318 (in Russian).

Filatov, V. A. (1969b). Calculation of magnetic field of a two-dimensional field of an arbitrary act. *Izv. Akad. Nauk SSSR. Fizika Zemli* **2**, 82–92 (in Russian).

Frank, F. and Mizes, R. (1937). *Differential and Integral Equations of Mathematical Physics.* ONTI, Moscow - Leningrad (in Russian).

Gakhov, F. D. (1963). *Boundary-Value Problems.* Fit. Math. Gos. Izd., Moscow (in Russian).

Glasko, V. B., Ostromogil'skii, A. Kh., and Filatov, V. G. (1970). On recovery of the depth and the form of a contact surface on the basis of the regularization method. *Journ. of Calc. Math. and Math. Phys.* **10**(5), 1292–1297 (in Russian).

Glasko, V. B., Litvinenko, O. K., Strakhov, V. N. *et al.*(1977). The Tikhonov regularization method in the modern prospecting geophysics. *Izv. Akad. Nauk SSSR. Fizika Zemli* **1**, 24–35 (in Russian).

Glasko, V. B. and Starostenko, V. I. (1976). The regularizing algorithm of solution of a system of nonlinear equations in inverse geophisic problems. *Izv. Akad. Nauk SSSR* **3**, 44–53 (in Russian).

Golizdra, G. Ya. (1966). *Singular Points of Analytic Extension of a Gravitational Field and their Connection with the Form of Perturbing Masses.* Add Chapters of Magnetic Prospecting Course. Nauka, Novosibirsk (in Russian).

Gurevich, Yu. M. (1982). *Methods of Calculation of Magnetic Fields Currents.* Sverdlovsk (in Russian).

Hachai, O. A. (1990). On solution of inverse problem for 3-D electromagnetic anomalies. *Izv. Akad. Nauk SSSR. Fizika Zemli* **2**, 55–59 (in Russian).

Henl, H., Maue, N. and Vestphal', K. (1964). *Diffraction Theory.* Mir, Moscow (in Rissian).

Himmelblau, D. (1975). *Applied Nonlinear Programming.* Mir, Moscow (in Russian).

Ivanov, V. K. (1956a). Integral equation of the inverse potential problem. *Dokl. Akad. Nauk SSSR* **105**(3), 400–412 (in Russian).

Ivanov, V. K. (1956b). On solvability of inverse problem of logarithmic potential in a finite form. *Dokl. Akad. Nauk SSSR* **106**(5), 1021–1024 (in Russian).

Ivanov, V. K. (1956c). Distribution of peculiarities of a potential and a spatial analog of the Polia theorem. *Math. Sbornik* **40**, 333–341 (in Russian).

Ivanov, V. K. (1963). On ill-posed problems. *Math. Sbornik* **61**, 211–223 (in Russian).

Ivanov, V. K. (1967). On integral Fregholm equations of the first kind. *Differential Equations* **3**(3), 410–421 (in Russian).

Ivanov, N. A. and Stadukhin, V. D. (1974). Application of MAM for determination of magnetic anomalies. *Prospecting and Protecting Nature* 142–148 (in Russian).

Kamenskaya, T. Ya. (1966). Determination of the cause of weak magnetic anomalies by means of observation of day variations of geomagnetic field. *Information of ONTI VIEMS* **3**, 18–24 (in Russian).

Kormil'tsev, V. V., Semenov, V. D., and Fedorov, I. M. (1975). On the charge method with magnetic field measurement, and its application. In: *Electroprospecting in the Sverdlovsk Domain.* Nauka, Sverdlovsk, p. 79 (in Russian).

Krylov, V. I. (1959). *Approximate Calculations of Integrals.* Fizmath. Gos. Izd., Moscow (in Russian).

Lavrentiev, M. M. (1956a). On the Cauchy problem for the Laplace equation. *Izv. Akad. Nauk SSSR. Ser. Mat.* **20**, 53–60 (in Russian).

Lavrentiev, M. M. (1956b). To the problem of an inverse problem of the potential theory. *Dokl. Akad. Nauk. SSSR* **106**, 861–864 (in Russian).

Lavrentiev, M. M. (1960). On integral equations of the first kind. *Dokl. Akad. Nauk SSSR* **133**, 1102–1105 (in Russian).

Lavrentiev, M. M. (1962). *On Certain Ill-Posed Problems of Mathematical Physics.* Nauka, Novosibirsk (in Russian).

Lavrentiev, M. M. (1965). A class of inverse problems for differential equations. *Dokl. Akad. Nauk SSSR* **160**(1), 33–35 (in Russian).

Makagonov, P. P. (1977). Skin-effect in the case of the step change of magnetic field. *Izv. Vusov, Geology and Prospecting* **7**, 134–137 (in Russian).

Markushevich, A. I. (1946). A boundary-value problem of theory of analytic functions. *Uch. Zap. Mosk. Gos. Univ.* **100**, 72–81 (in Russian).

Martyshko, P. S. (1981). An algorithm of solution of direct and inverse two-dimensional problems of magnetic prospecting taking account of dismagnetization effect. In: *Geology and Poleznye Iskopaemye Urala.* Nauka, Sverdlovsk, pp. 64–65 (in Russian).

Martyshko, P. S. (1982a). Solution of problems of the method of artifitial magnetization in two and three-dimensional cases. In: *Theory and Practice of Interpetation of Gravimetry and Magnetic Fields in the USSR.* Kiev, pp. 198–204 (in Russian).

Martyshko, P. S. (1982b). *Some Problems of Theory and Algorithms of Solution of the Method of Artifitial Magnetization.* Sverdlovsk (in Russian).

Martyshko, P. S. (1983). Solution of the direct and inverse three-dimensional problems of the method of artifitial magnetization in the parameter classes. *Izv. Akad. Nauk SSSR. Fizika Zemli* **3**, 52–57 (in Russian).

Martyshko, P. S. (1986a). Solution of an inverse problem of electroprospecting for a direct current for arbitrary classes of potentials. *Izv. Akad. Nauk SSSR. Fizika Zemli* **1**, 87–92 (in Russian).

Martyshko, P. S. (1986b). Solution of an inverse problem for magnetic field of scattering currents. In: *Geology and Poleznye Iskopaemye Urala.* Sverdlovsk, pp. 55–56 (in Russian).

Martyshko, P. S. (1988). Integro-differential equation for an inverse problems for magnetic field of scattering currents. In: *Methods of Interpret and Math. Model of Geophys. Fields.* Sverdlovsk, pp. 24–27 (in Russian).

Martyshko, P. S. (1989). Equation for inverse problems with the operator given explicity for variable electromagnetic fields. In: *Appl. of Math. Meth. and Computers when Processing Information in Geology.* Chelyabinsk, p. 102 (in Russian).

Martyshko, P. S. (1990a). Solution of inverse problem for variable electromagnetic fields. In: *Electr. Induction in the Earth Surface.* Moscow, p. 89 (in Russian).

Martyshko, P. S. (1990b). Integro-differential equations for variable electromagnetic fields. *Izv. Akad. Nauk SSSR. Fizika Zemli* **5**, 55–62 (in Russian).

Martyshko, P. S. (1991). Explicit equations for inverse problems for electromagnetic fields. In: *Tez. Dokl. III Sci-Tech. Conf. in Geotomography.* Sverdlovsk, p. 80 (in Russian).

Martyshko, P. S. (1992). Equation of inverse problem for the wave electromagnetic field. In: *Theory and Practice of Electromagnetic Methods of Geophys. Researches.* Ekaterinburg, pp. 3–6 (in Russian).

Martyshko, P. S. (1993). On solution of the inverse problem of the charge method. *Izv. Ros. Akad. Nauk. Fizika Zemli* **5** (in Russian).

Martyshko, P. S. and Prutkin, I. L. (1982). Solution of the direct and inverse magnetic prospecting problem. *Geophys. Journ.* **4**(6), 39–49 (in Russian).

Martyshko, P. S., Tsirul'skii, A. V., and Gurevich, Yu. M. (1986). On solvability of research of magnetic anomalies given the data of the artifitial magnetization method. In: *Glubinnoye Stroyenie i Poleznye Iskopaemye Vostoka SSSR.* Vladivostok, pp. 54–70 (in Russian).

Martyshko, P. S., Gurevich, Yu. M., and Turanov, V. N. (1988). Interpretation of square probing data of MAM. In: *Meth. of Interpr. and Math. Model of Geophys. Fields.* Sverdlovsk, pp. 33–39 (in Russian).

Martyshko, P. S. and Pyankov, V. A. (1989). On uniqueness of solution of the inverse problem of the theory of potential in the method of magnetization by rotating field. (Sq-variations). In: *Problems of Theory and Results of Application of Methods of Interpretation of Geophys. Fields.* Sverdlovsk, pp. 13–19 (in Russian).

Mayergoyz, I. D. (1979). *Iterative Methods of Calculation of Statistic Fields in Inhomogeneous Anisotropy and Nonlinear Media.* Kiev (in Russian).

Mikhailov, L. G. (1963). General problem of conjugation of analytic functions and its application. *Izv. Akad. Nauk SSSR. Ser. Mat.* **27**(5), 969–993 (in Russian).

Mors, F. M. and Feshbakh, G. (1958). *Methods of Theoretical Physics.* Izd. Inostr. Lit., Moscow (in Russian).

Muskhelishvili, N. I. (1962). *Singular Integral Equations.* Fizmatgiz, Moscow (in Russian).

Nadirashvili, N. S. (1982). On classes of gravimetrically equivalent bodies. In: *Theory and Practice of Interpr. of Gravitat. Anomalies.* Moscow, pp. 15–25 (in Russian).

Nikonova, F. I. and Tsirul'skii, A. V. (1978). Interpretation of gravimagnetic anomalies on the basis of the class of potentials for which the problem is soluble in a finite from. *Izv. Akad. Nauk SSSR. Fizika Zemli* **2**, 49–60 (in Russian).

Ortega, Dj. and Reinbold, V. (1975). *Iterative Methods for Solution of Nonlinear System of Equations With Many Unknowns.* Mir, Moscow (in Russian).

Petrov, A. A. and Fedorov, A. N. (1988). Solution of inverse problem of electroprospecting by the methods of direct currect and the polarization brought about. *Izv. Akad. Nauk SSSR, Fizika Zemli.* **11**, 60–64, Moscow (in Russian).

Petrov, A. A. and Fedorov, A. N. (1989). Interpretation of data of electroprospecting by direct current in the case of irregular surface. *Izv. Akad. Nauk SSSR. Fizika Zemli* **12**, 84–88 (in Russian).

Rapoport, I. M. (1940). On some sufficient conditions of uniqueness of solution of the inverse potential problem. *Dokl. Akad. Nauk SSSR* **5**, 23–30 (in Russian).

Rozenbrok, Kh. and Story, S. (1968). *Calculation Methods.* Mir, Moscow (in Russian).

Semenov, V. D., Kormil'tsev, V. V., and Gurevich, Yu. M. (1977). The charge Method with Measurement of magnetic field strength in prospecting and exploring pyrites deposits. In: *Electric Research in Prospecting and Exploring Pyrites Deposits.* Sverdlovsk, pp. 14–20 (in Russian).

Sharma, P. V. (1966). Rapid computation of magnetic anomalies and demagnetization effects caused by bodies of arbitrary shape. *Pure Appl. Geophys.* **64**, 89–109.

Sobolev, V. V. and Belogolov, V. T. (1968). Calculation of intensity of magnetization of rock stuff in homogeneous magnetization fields taking account of demagnetization. *Trudy SNIIGGIMS* **73**, 33–39 (in Russian).

Sobolev, S. L. (1962). *Equations of Mathematical Physics.* Nauka, Moscow (in Russian).

Sobolev, O.A. (1960). The problem of dividing the magnetic anomalies. *Geology and Geophysics* **9**, 67–77 (in Russian).

Sretenskii, L. N. (1946). *Theory of the Newton Potential.* Gos. Tekh. Izdat., Moscow (in Russian).

Stadukhin, V. D. (1962). Determination of magnetic properties of the rock stuff in natural position by measured strength of magnetic field in the center of the square. *Trudy Inst. Geology. Sverdlovsk* **2**, 85–90 (in Russian).

Starostenko, V. I. and Oganesyan, S. M. (1977). Stable operator processes and their application in the geophysical problems. *Izv. Akad. Nauk SSSR. Fizika Zemli* **5**, 61–75 (in Russian).

Strakhov, V. N. (1963a). On analytic extension of electric fields applying in certain methods of electroprospecting with direct current. *Izv. Akad. Nauk SSSR. Ser. Geofiz.* **3**, 416–419 (in Russian).

Strakhov, V. N. (1963b). On analytic extension of electric fields in conductive half-space. *Izv. Akad. Nauk SSSR. Ser. Geofiz.* **3**, 419–431 (in Russian).

Strakhov, V. N. (1967). On solution of ill-posed problems of magnetometry and gravimetry which reduced to the integral equations of the convolution type. *Izv. Akad. Nauk SSSR. Fizika Zemli* **4**, 36–55; **5**, 33–54 (in Russian).

Strakhov, V. N. (1969). Theory of approximate solution of linear ill-posed problems in the Gilbert space and its application in prospecting of geophysics. *Izv. Akad. Nauk SSSR. Fizika Zemli* **8**, 30–54; **9**, 64–97, Moscow (in Russian).

Strakhov, V. N. (1970). Some problems of the plane inverse problem of the magnetic potential. *Izv. Akad. Nauk SSSR. Fizika Zemli* **9**, 58–71 (in Russian).

Strakhov, V. N. (1973). Some examples of equivalence and weak uniqueness in the plane inverse problem of potential. *Izv. Akad. Nauk SSSR. Ser. Geofiz.* **5**, 39–52 (in Russian).

Strakhov, V. N. (1974). Functional equations of the plane inverse problem of potential and numerical approximate solutions of it. *Dokl. Akad. Nauk SSSR. Ser. Geofiz.* **213**(1), 863–866 (in Russian).

Strakhov, V.N. (1977). To the theory of the method of artifitial magnetization. *Izv. Akad. Nauk SSSR. Fizika Zemli* **7**, 56–74 (in Russian).

Strakhov, V. N. and Blokh, Yu. I. (1977a). Influence of demagnetization for determination of harmonic moments of two-dimensional bodies of an arbitrary from. *Razved. Geofizika* **75**, 77–81 (in Russian).

Strakhov, V. N. and Blokh, Yu. I. (1977b). Approximative way of determination of harmonic moments of two-dimensional magnetized objects. *Razved. Geofizika* **75**, 82–86 (in Russian).

Svetov, B. S. (1973). *Theory, Methods and Interpretation of Low-Frequency Inductive Electroprospecting Data.* Nedra, Moscow (in Russian).

Svetov, B. S. (1984). *Electrodynamic Basis of Quasistationary Geoelectrics.* IZMI RAN, Moscow (in Russian).

Svetov, B. S. and Gubatenko, V. P. (1982). On scalarization of the Maxwell equations in inhomogeneous media. *Izv. Akad. Nauk SSSR, Fizika Zemli.* **11**, 63–71, Moscow (in Russian).

Svetov, B. S. and Gubatenko, V. P. (1988). *Analytic Solution of Electrodynamic Problems.* Nauka, Moscow (in Russian).

Stretton, Dj. (1948). *Electromagnetic Theory.* Moscow (in Russian).

Talwani, M. (1965). Computation with the help of arbitrary shape. *Geophys.* **30**, 797–808.

Tikhonov, A. N. (1943). On stability of inverse problems. *Dokl. Akad. Nauk SSSR* **39**(5), 501–504 (in Russian).

Tikhonov, A. N. (1963). On solution of ill-posed problems. *Dokl. Akad. Nauk SSSR* **151**(3), 791–794 (in Russian).

Tikhonov, A. N. and Arsenin, V. Ya. (1979). *Methods of Solution of Ill-Posed Problems.* Nauka, Moscow (in Russian).

Tikhonov, A. N. and Glasko, V. B. (1965). Application of the regularization methods in nonlinear problems. *Journ. of Calc. Math. and Math. Phys.* **5**(3), 463–473 (in Russian).

Tripp, A. C. and Hohmann, G. W. (1993). Three-dimensional electromagnetic cross-well inversion. *IEEE Trans. Geosc. Remote Sensing* **31**, 121–126.

Tsirul'skii, A. V. (1963). On some properties of the complex logarithmic potential of a homogeneous domain. *Izv. Akad. Nauk SSSR. Fizika Zemli* **7**, 1072–1075 (in Russian).

Tsirul'skii, A. V. (1969). On uniqueness of solution of the inverse problem of the theory of potential. *Izv. Akad. Nauk SSSR. Fizika Zemli* **6**, 60–65 (in Russian).

Tsirul'skii, A. V. (1972). On uniqueness of solution of the inverse problem of the theory of potential in the method of artifitial magnetization. *Izv. Akad. Nauk SSSR. Fizika Zemli* **9**, 64–67 (in Russain).

Tsirul'skii, A. V. (1974a). On solution of the direct and inverse problems of gravity prospecting. *Izv. Akad. Nauk SSSR. Fizika Zemli* **7**, 81–87 (in Russain).

Tsirul'skii, A. V. (1974b). To the theory of the method of artifitial magnetization in the two-dimensional case. *Izv. Akad. Nauk SSSR. Fizika Zemli* **9**, 70–76, Moscow (in Russian).

Tsirul'skii, A. V. (1975). To the problem of solution of the direct electroprospecting problem. In: *Electric Prospecting.* Sverdlovsk, pp. 89–95 (in Russian).

Tsirul'skii, A. V. (1976). *Inverse Problems of Logarithmic Pitential.* Diss. Dok. Fiz.-Mat. Nauk, Sverdlovsk (in Russian).

Tsirul'skii, A. V. (1990). *The Complex Functions in the Theory and Methods of Potential Geophysical Fields.* Sverdlovsk (in Russian).

Tsirul'skii, A. V. and Nikonova, F. I. (1975). On solvability of the inverse problem of the logarithmic potential in a finite form. *Izv. Akad. Nauk SSSR. Fizika Zemli* **5**, 37–46 (in Russian).

Tsirul'skii, A. V. and Martyshko, P. S. (1979). On account of dismagnetic effect in the magnetic prospecting problem. *Izv. Akad. Nauk SSSR. Fizika Zemli* **3**, 49–56 (in Russian).

Tsirul'skii, A. V. and Prutkin, I. L. (1981). On solution of the inverse gravimetry problem for arbitrary classes of two- and three-dimensional potentials. *Izv. Akad. Nauk SSSR. Fizika Zemli* **11**, 54–61 (in Russian).

Undezenkov, B. A. (1978). On efficiency of magneto-variational method when research and prospecting of magnetic stuff. *Magnetic mestorozdenie of Ural.* Sverdlovsk, pp. 144–138 (in Russian).

Vargova, E. (1977). Solution of the direct inhomogeneous magnetic problem by means of analytical four-vectors. *Contribs Geofys. Inst. Slovak Acad. Sci.* **8**, 763–767.

Vekua, I. N. (1959). *Generalized Analytic Functions.* Fizmatgiz, Moscow (in Russian).

Vogel, A. (1963). The application of electronic computers to the calculation of effective magnetization. *Geophys. Prosp.* **11**, 51–58.

Voskoboynikov, G. M. (1973). On calculation of stationary electromagnetic fields in piecewise-homogeneous media. *Izv. Akad. Nauk SSSR. Fizika Zemli* **3**, 63–75 (in Russian).

Weidelt, P. (1975). Electromagnetic induction in three-dimensional structures. *J. Geophys.* **416**(1), 85–109.

Xiong, Z. and Kirsch, A. (1992). Three-dimensional earth conductivity inversion. *J. Comp. Appl. Math.* **42**, 109–121.

Yanovskii, B. M. (1978). *Earth Magnetizm.* Gos. Knot. Leningrad, Leningrad (in Russian).

Zakharov, E. V. and Il'in, I. V. (1970). Integral representations of electromagnetic fields in the inhomogeneus layered medium. *Izv. Akad. Nauk SSSR. Fizika Zemli* **8**, 62–71 (in Russian).

Zamorev, A. A. (1941). The inverse two-dimensional problem of the theory of potential. *Dokl. Akad. Nauk SSSR* **31**(9), 262–265 (in Russain).

Zhdanov, M. S. (1973). Division of variable electromagnetic Earth fields. *Izv. Akad. Nauk SSSR. Fizika Zemli* **6**, 43–54 (in Russian).

Zhdanov, M. S. (1975). Problems of the theory of interpretation of the depth electromagnetic anomalies on the basis of the methods of analytic extension. *Izv. Akad. Nauk SSSR. Fizika Zemli* **7**, 59–73 (in Russian).

Zhdanov, M. S. (1984). *Analogs of the Type Cauchy Integrals in the Theory of Geophysical Fields.* Nauka, Moscow (in Russian).

Zhdanov, M. S. and Frenkel', M. A. (1983). Method of electromagnetic migration for solution of inverse geoelectric problems. *Dokl. Akad. Nauk SSSR* **271**(3), 589–594 (in Russian).

Bei Fragen zur Produktsicherheit wenden Sie sich bitte an:
If you have any questions regarding product safety,
please contact:

Walter de Gruyter GmbH
Genthiner Straße 13
10785 Berlin
productsafety@degruyterbrill.com